Studies in Systems, Decision and Control

Volume 521

The series "Studies in Systems, Decision and Control" (SSDC) covers both new developments and advances, as well as the state of the art, in the various areas of broadly perceived systems, decision making and control–quickly, up to date and with a high quality. The intent is to cover the theory, applications, and perspectives on the state of the art and future developments relevant to systems, decision making, control, complex processes and related areas, as embedded in the fields of engineering, computer science, physics, economics, social and life sciences, as well as the paradigms and methodologies behind them. The series contains monographs, textbooks, lecture notes and edited volumes in systems, decision making and control spanning the areas of Cyber-Physical Systems, Autonomous Systems, Sensor Networks, Control Systems, Energy Systems, Automotive Systems, Biological Systems, Vehicular Networking and Connected Vehicles, Aerospace Systems, Automation, Manufacturing, Smart Grids, Nonlinear Systems, Power Systems, Robotics, Social Systems, Economic Systems and other. Of particular value to both the contributors and the readership are the short publication timeframe and the world-wide distribution and exposure which enable both a wide and rapid dissemination of research output.

Indexed by SCOPUS, DBLP, WTI Frankfurt eG, zbMATH, SCImago.

All books published in the series are submitted for consideration in Web of Science.

Axaykumar Mehta · Akash Modi

Robust Sliding Mode Protocols for Formation of Quadcopter Swarm

Springer

Axaykumar Mehta
Department of Electrical and Computer Science Engineering
Institute of Infrastructure Technology Research and Management
Ahmedabad, India

Akash Modi
Department of Electrical and Computer Science Engineering
Institute of Infrastructure Technology Research and Management
Ahmedabad, India

ISSN 2198-4182 ISSN 2198-4190 (electronic)
Studies in Systems, Decision and Control
ISBN 978-981-99-9725-1 ISBN 978-981-99-9726-8 (eBook)
https://doi.org/10.1007/978-981-99-9726-8

This Springer imprint is published by the registered company Springer Nature Singapore Pte Ltd.
The registered company address is: 152 Beach Road, #21-01/04 Gateway East, Singapore 189721, Singapore

Paper in this product is recyclable.

Axaykumar Mehta dedicated this book to his mother Jayaben, wife Dr. Hema, and loving sons Siddharth, and Yatharth

and

Akash Modi dedicated this book to his father Kalpeshkumar, mother Tejalben, and beloved Prachi, and Krish

Preface

In recent years, advancements in robotics, artificial intelligence, autonomous systems, and IoTs have ushered in a new era of possibilities in various research fields, including aerial vehicle systems. Among these innovations, the Quadcopter swarm has emerged as a testament to the astonishing research areas due to its civilian and military applications. A Quadcopter, a type of unmanned aerial vehicle (UAV) characterized by its four rotors, has rapidly transitioned from a recreational gadget to a versatile tool with an ever-expanding range of applications. However, it is the notion of Quadcopter swarm that has captivated the imagination of researchers, engineers, and the wider public alike. Imagine a coordinated ensemble of autonomous Quadcopters, seamlessly collaborating to achieve tasks that were once deemed insurmountable for a single drone. This concept has not only sparked interest but has also paved the way for a new era of possibilities. A Quadcopter swarm refers to a group of multiple Quadcopters that operate together in a coordinated manner. By leveraging the power of multiple Quadcopters working together, swarm technology offers several advantages over single drone operations, including increased efficiency, improved redundancy, and enhanced capabilities. However, it is required to design a robust formation protocol to ensure the efficient swarm behavior. The goal of formation protocol is to maintain a specific spatial arrangement between the Quadcopters, while also ensuring that each Quadcopter avoids collisions with other Quadcopters and obstacles in the environment. Therefore, it is believed that the formation of multiple low-cost Quadcopters is more effective in accomplishing the goal than a single high-tech Quadcopter.

On the other hand, the Multi-Agent System (MAS) concept has been exploited widely for controlling and synchronizing multiple systems like robotic arms, helicopter systems, and many more. In each of these applications, the agents in the system work together to achieve a common goal, such as maximizing efficiency, minimizing congestion, or optimizing profits. MAS is a computational system made up of numerous autonomous agents also termed as systems that collaborate to accomplish shared or unique objectives. MAS has grown in popularity due to its capability to address complicated problems that are challenging for single agents to manage. The primary concept behind the cooperative control of the MAS is to create a distributed controller for each agent using information regarding the surrounding environment in order to achieve certain objectives. In literature, many algorithms have been proposed for cooperative controls such as formation, flocking, rendezvous, and consensus. However, in the application of Quadcopter swarm, formation using MAS is the popular algorithm. The research of the formation protocol can be classified into two parts. The first part is working on the formation without a leader and the second one formation with the leader. In case of the leader-follower protocol using MAS, there are mainly two types of controllers designed, namely, (1) primary controller and (2) distributed or secondary controller. The primary controller stands for the individual controller of the Quadcopter to get the desired setpoint or to maintain the desired output. However, distributed controller is designed for maintaining the appropriate distance between the neighborhood Quadcopters which can achieve the formation as per the given pattern. Many researchers have contributed to the field of the formation of multiple Quadcopters. Few of them have designed the formation control of Quadcopter swarm using conventional controllers such as proportional-derivative (PD), proportional-integral-derivative (PID), model-predictive control (MPC), and linear quadratic regulator (LQR). However, the conventional control techniques could not be adequately efficient to maintain the formation under any environmental uncertainty. So as a result, researchers have chosen the sliding mode control (SMC) having the advantages of robustness, finite-time convergence, and simplicity in design.

This monograph presents several such robust sliding mode protocols for the formation of Quadcopter swarm which achieves the formation within the finite-time in continuous-time domain and Discrete-time domain. A fixed graph theory is used to represent the communication between the Quadcopters and leader-follower MAS is used to achieve the formation. For this work, 6-DOF (Degrees of Freedom) Quadcopter model has been used. First, the continuous-time primary control using the linear quadratic regulator (LQR) technique for the linearized model of the Quadcopter is designed and then a distributed controller using the second-order sliding mode (SOSM) super-twisting configured with the fixed and undirected graph topology is designed to achieve the Quadcopter swarm formation. The Lyapunov function is used to endorse the global stability of the swarm. However, the controllers designed using the linear models are not performing well. Hence, the non-linear model of the Quadcopter is used to design the primary controller using the SMC technique. Further, a distributed SOSM super-twisting protocol was designed using the non-linear model of Quadcopter swarm to achieve the formation.

From the literature survey on Quadcopter swarm, it is observed that the majority of protocols are designed in the continuous-time domain. However, nowadays in the application of Quadcopters swarm, the digital processors and the digital communication are used to leverage the Digital implementation. It is always challenging to implement the continuous-time SMC for formation using a digital processor due to high-frequency control efforts for maintaining the formation among the agents.

This is the reason, why we resort ourselves to design the formation protocols in the discrete-time domain. Like continuous-time protocols, first a primary controller is designed using the discrete-time LQR technique and then a discrete-time sliding mode (DSM) formation protocol using the power rate reaching law is designed for the Quadcopter swarm. The performance of the designed protocol is compared with the other classical controllers designed using Gao's reaching law. Also, the analysis for the number of steps required for the global formation of DMAS along with the Lyapunov stability is achieved. While analyzing the results, it is observed that there is a significant amount of chattering band present in the control effort. So, in order to reduce the chattering band, a discrete-time SOSM super-twisting formation protocol is designed to achieve the formation. The required number of steps for global consensus is also derived. Finally, a DDSM protocol using the exponential reaching law for the formation of Quadcopter swarm. The exponential reaching law is also inherently more robust and provides the minimum chattering band than the other controllers. In this case, also the number of steps required for the global formation along with the global stability of DMAS is derived using the Lyapunov function.

Summarizing, this monograph contributes to the robust sliding mode protocols for the formation of leader-following Quadcopter swarm using the MAS concept in continuous-time and discrete-time domain. The above-mentioned work is divided into the following chapters.

- Chapter 1 presents a brief introduction about the 6-DOF (Degree of Freedom) Quadcopter, and the literature review on the formation protocol for Quadcopter swarm in the continuous-time and discrete-time domain.
- Chapter 2 briefs the detailed information about the MAS concept, graph theory, and the SMC technique which are used to derive the formation protocols in the continuous-time and discrete-time domain.
- Chapter 3 presents the design of the continuous-time sliding mode formation protocols for Quadcopter swarm using the linear and non-linear model of Quadcopter.
- Chapter 4 presents the design of a discrete-time sliding mode (DSM) protocol for the formation of Quadcopter swarm. The protocol is designed using the power rate reaching law and compared with the classical formation protocol developed using Gao's reaching law. The chapter also includes the calculation of number of steps required to achieve the formation.
- Chapter 5 presents the design of discrete-time second-order sliding mode (DSOSM) protocol using discrete-time super-twisting (DST) protocol for the formation of Quadcopter swarm. The protocol reduces the Quasi sliding mode band (QSMB) and faster the convergence time of the formation of DMAS. Further to show the efficacy of the proposed protocol, the results are compared with the first-order DSM protocols designed using Gao's reaching law and Bartoszewicz's reaching law.

- Chapter 6 presents the design of DSM protocol using a novel exponential reaching law for the formation of Quadcopter swarm along with the required number of steps to achieve the formation and the Lyapunov stability analysis. The protocol is also compared with the other classical protocol designed using Gao's reaching law.
- Chapter 7 concludes with a summary of the main findings and the future research possibilities.

Ahmedabad, India
October 2023

Axaykumar Mehta
Akash Modi

Contents

1 Introduction 1
1.1 Overview of Quadcopter Swarm 1
1.2 Literature Review 5
1.2.1 Technical Approaches for Formation of Quadcopter Swarm 6
1.3 Organization of Book 7
References 8

2 Preliminaries of Multi-agent System, Graph Theory, and Sliding Mode Control 11
2.1 Overview of Multi-agent System 11
2.2 Preliminaries of Algebraic Graph Theory 14
2.3 Review of Sliding Mode Control 15
2.3.1 Continuous-Time Sliding Mode Control 17
2.3.2 Discrete-Time Sliding Mode Control 20
2.4 Conclusion 22
References 22

3 Continuous-Time Sliding Mode Protocol for Formation of Quadcopter Swarm 25
3.1 System Description of Quadcopter 25
3.2 Distributed Continuous-Time Sliding Mode (DCSM) Formation Protocol for Linear Model of Quadcopter Swarm 26
3.2.1 Linear Model of Quadcopter 27
3.2.2 Continuous-Time Primary Control for Linear Model of Quadcopter 30
3.2.3 Simulation Results of Continuous-Time Primary Control for Linear Model of Quadcopter 31
3.2.4 Graph Theory 34
3.2.5 Multi-agent System (MAS) 34

3.2.6 DCSM Formation Protocol for Linear Model of Quadcopter Swarm Using Constant-Proportional Reaching Law ... 36
3.2.7 Simulation Results of DCSM Formation Protocol Using Constant-Proportional Reaching Law ... 39
3.2.8 Distributed Continuous-Time Second-Order Sliding Mode (DCSOSM) Super-Twisting (ST) Formation Protocol for Linear Model of Quadcopter Swarm ... 43
3.2.9 Simulation Results of DCSOSM ST Formation Protocol ... 45
3.3 DCSM Formation Protocol for Non-linear Model of Quadcopter Swarm ... 49
3.3.1 Non-linear Model of Quadcopter ... 49
3.3.2 Continuous-Time Primary Control for Non-linear Model of Quadcopter ... 50
3.3.3 Simulation Results of Continuous-Time Primary Control for Non-linear Model of Quadcopter ... 54
3.3.4 DCSOSM ST Formation Protocol for Non-linear Model of Quadcopter Swarm ... 55
3.3.5 Simulation Results of Non-linear model of Quadcopter using DCSOSM ST formation Protocol ... 59
3.4 Conclusion ... 60
References ... 63

4 Discrete-Time Sliding Mode Protocol for Formation of Quadcopter Swarm ... 65
4.1 System Description and Mathematical Modeling of Quadcopter ... 65
4.2 Discrete-Time Primary Control of Quadcopter ... 71
4.2.1 Simulation Results of Discrete-Time Primary Control of Quadcopter ... 72
4.3 Distributed Discrete-Time Sliding Mode (DDSM) Protocol for Formation of Quadcopter Swarm ... 73
4.3.1 Discrete Multi-agent System (DMAS) ... 73
4.3.2 DDSM Formation Protocol for Quadcopter Swarm Using Power Rate Reaching Law ... 79
4.3.3 The Analysis of Number of Steps Required to Achieve the Formation of Quadcopter Swarm with DDSM Protocol Using Power Rate Reaching Law ... 81
4.3.4 Simulation Results of DDSM Formation Protocol Using Power Rate Reaching Law ... 85
4.4 Comparative Analysis of DDSM Using Power Rate Reaching Law with DDSM Using Gao's Reaching Law ... 89

4.4.1 The Analysis of Number of Steps Required to Achieve the Formation of Quadcopter Swarm with DDSM Protocol Using Gao's Reaching Law ... 91
4.4.2 Simulation Results of DDSM Formation Protocol Using Gao's Reaching Law ... 95
4.5 Conclusion ... 99
References ... 99

5 Discrete-Time Second-Order Sliding Mode Protocol for Formation of Quadcopter Swarm ... 101
5.1 Introduction to Discrete Second-Order Sliding Model Protocol ... 101
5.2 Distributed Discrete-Time Second-Order Sliding Mode (DDSOSM) Super-Twisting (ST) Protocol for Formation of Quadcopter Swarm ... 102
5.2.1 Simulation Results of DDSOSM ST Formation Protocol ... 106
5.3 Comparative Analysis of DDSOSM ST Protocol with DDSM Protocol Using Bartoszewicz's Reaching Law ... 110
5.3.1 Simulation Results of DDSM Formation Protocol Using Bartoszewicz's Reaching Law ... 114
5.4 Comparative Analysis of DDSOSM ST Protocol with DDSM Protocol Using Gao's Reaching Law ... 114
5.4.1 Simulation Results of DDSM Formation Protocol Using Gao's Reaching Law ... 115
5.5 Conclusion ... 116
References ... 121

6 Discrete-Time Sliding Mode Protocol Using Exponential Reaching Law for Formation of Quadcopter Swarm ... 123
6.1 Introduction to Discrete-Time Exponential Sliding Model Protocol ... 123
6.2 DDSM Formation Protocol for Quadcopter Swarm Using Exponential Reaching Law ... 124
6.2.1 Analysis of Number of Steps Required for Formation of Quadcopter Swarm with DDSM Using Exponential Reaching Law ... 127
6.2.2 Simulation Results of DDSM Formation Protocol Using Exponential Reaching Law ... 131
6.3 Comparative Analysis of DDSM Using Exponential Reaching Law with DDSM Protocol Using Gao's Reaching Law ... 137
6.4 Conclusion ... 142
References ... 142

7 Concluding Remarks and Future Scope ... 145

About the Authors

Axaykumar Mehta received B.E. Electrical Engineering (1996), degree from LD College of Engineering Ahmedabad, M.Tech. in Control Systems (2002) from IIT Kharagpur, and Ph.D. (2009) from IIT Bombay in systems and control engineering. Currently, he is acting as a Professor in Electrical and Computer Science Engineering Department and also holding the position of an Associate Dean Academic Programs at the Institute of Infrastructure, Technology, Research and Management (IITRAM), Ahmedabad, India. Priorly, he has acted as a Director at Gujarat Power Engineering and Research Institute, Gujarat, India, as an Associate faculty at Indian Institute of Technology Gandhinagar, as visiting professor at Institute of Automatic Control, Lodz University of Technology Poland, and a Research Associate at University of Cagliari, Italy. His research interests are sliding mode control, networked control and control of multi-agent systems and their applications to robotics, aerial vehicles, microgrid etc... He has published several research papers in reputed peer reviewed journals and conference proceedings. He has contributed four monographs, one editorial book, and three conference proceedings on the research topics with Springer Nature Singapore. He is member of Technical Committee on Variable Structure and Sliding Mode Control under the aegis of IEEE Control System Society, Fellow member—Institute of Engineers (I), Senior member—IEEE, IEEE Industrial Electronics Society (IES), IEEE Control System Society, Life member of Indian Society for Technical Education (ISTE), and Systems Society of India (SSI). He is also recognized as a Chartered Engineer in Electrical Engineering by the Institute of Engineers (I).

Akash Modi received his B.Tech. (2021) in Electrical and Computer Science Engineering, and M.Tech. (2023) in Electrical Infrastructure from Institute of Infrastructure, Technology, Research and Management (IITRAM), Ahmedabad, India. His current research interests are distributed sliding mode control, and control of multi-agent system and their applications. He has also worked in the area of design the micro-strip patch antenna, multi-layer antenna, and re-configurable array antenna for commercial and satellite applications. He has published a several research papers on the research topic with reputed publishers. He has received an Undergraduate Research Award-2 (URA-2) from the IITRAM in 2021.

Acronyms

Abbreviations

BLDC	Brushless DC
CSMC	Continuous-time Sliding Mode Control
DAI	Distributed Artificial Intelligence
DCSM	Distributed Continuous-time Sliding Mode
DCSOSM	Distributed Continuous-time Second-Order Sliding Mode
DCSOSM	Distributed Discrete-time Second-Order Sliding Mode
DDSM	Distributed Discrete-time Sliding Mode
DMAS	Discrete Multi-Agent System
DOF	Degrees of Freedom
DPS	Distributed Problem Solving
DSMC	Discrete-time Sliding Mode Control
DSTA	Discrete Super-Twisting Algorithm
ESCs	Electronic Speed Controllers
HOSMC	Higher Order Sliding Mode Control
ISMC	Integral Sliding Mode Control
LQR	Linear Quadratic Regulator
MAS	Multi-Agent System
MPC	Model Predictive Control
SMC	Sliding Mode Control
STA	Super-Twisting Algorithm
UAV	Unmanned Aerial Vehicle
VTOL	Vertical and Takeoff Landing

Symbols

x(t)	State vector of individual system in continuous-time domain
y(t)	Output of individual system in continuous-time domain
u(t)	Input vector of individual system in continuous-time domain
x(k)	State vector of individual system in discrete-time domain
y(k)	Output of individual system in discrete-time domain
u(k)	Input vector of individual system in discrete-time domain
A(t)	System matrix in continuous-time domain
B(t)	Pant matrix in continuous-time domain
E(k)	System matrix in discrete-time domain
F(k)	Plant matrix in discrete-time domain
x	Linear position of Quadcopter in x direction
$\dot{x}$	Linear velocity of Quadcopter in x direction
$\ddot{x}$	Linear acceleration of Quadcopter in x direction
$\dddot{x}$	Linear jerk of Quadcopter in x direction
y	Linear position of Quadcopter in y direction
$\dot{y}$	Linear velocity of Quadcopter in y direction
$\ddot{y}$	Linear acceleration of Quadcopter in y direction
$\dddot{y}$	Linear jerk of Quadcopter in y direction
z	Linear position of Quadcopter in z direction
$\dot{z}$	Linear velocity of Quadcopter in z direction
$\ddot{z}$	Linear acceleration of Quadcopter in z direction
$\dddot{z}$	Linear jerk of Quadcopter in z direction
θ	Angular position of Quadcopter in x direction
$\dot{\theta}$	Angular velocity of Quadcopter in x direction
$\ddot{\theta}$	Angular acceleration of Quadcopter in x direction
$\dddot{\theta}$	Angular jerk of Quadcopter in x direction
ϕ	Angular position of Quadcopter in y direction
$\dot{\phi}$	Angular velocity of Quadcopter in y direction
$\ddot{\phi}$	Angular acceleration of Quadcopter in y direction
$\dddot{\phi}$	Angular jerk of Quadcopter in y direction
ψ	Angular position of Quadcopter in z direction
$\dot{\psi}$	Angular velocity of Quadcopter in z direction
$\ddot{\psi}$	Angular acceleration of Quadcopter in z direction
$\dddot{\psi}$	Angular jerk of Quadcopter in z direction
Q, R	Weighted matrices of LQR controller
K_{Cont}	Feedback gain of primary controller in continuous-time domain
$\otimes$	Kronecker product
$\bar{A}$	Global system matrix in continuous-time domain
$\bar{B}$	Global plant matrix in continuous-time domain
δ(t)	Local formation error continuous-time domain
sgn	Signum function
$\mathbb{E}$(t)	Global formation error continuous-time domain

d_{ij}	Local distance vector
$\bar{d}_{ij}$	Global distance vector
$\mathbb{S}_i(t)$	Sliding surface of *ith* agent in continuous-time domain
$\bar{\mathbb{S}}(t)$	Global sliding surface in continuous-time domain
Υ_i	Sliding mode gain of *ith* agent in continuous-time domain
$\bar{\Upsilon}$	Global sliding mode gain in continuous-time domain
Λ_1, Λ_2	Local user defined constants
$\bar{\Lambda}_1, \bar{\Lambda}_2$	Global user defined constants
$\bar{\beta}$	Stability condition for global formation
$\mathcal{V}(t)$	Lyapunov function for stability analysis
λ_i	User defined constant for *ith* agent continuous-time domain
$\bar{E}(k)$	Global system matrix in discrete-time domain
$\bar{F}(k)$	Global plant matrix in discrete-time domain
K_{dt}	Feedback gain of primary controller in discrete-time domain
τ	Sampling time
$\mathbb{S}_i(k)$	Sliding surface of *ith* agent in discrete-time domain
$\bar{\mathbb{S}}(k)$	Global sliding surface in discrete-time domain
Υ_d	Sliding mode gain of *ith* agent in discrete-time domain
$\bar{\Upsilon}_d$	Global sliding mode gain in discrete-time domain
λ, μ	Global user defined constants
Δ_i	Quasi sliding mode band
$\widehat{k}$	Total number of steps required for formation
κ_{1i}, κ_{2i}	Local user defined constant for DDSOSM ST protocol
κ_1, κ_2	Global user defined constant for DDSOSM ST protocol
$q, \theta, \gamma, \beta, \delta$	Local user defined constant for DDSM using exponential reaching law
$\bar{q}, \bar{\theta}, \bar{\gamma}, \bar{\beta}, \bar{\delta}$	Global user defined constant for DDSM using exponential reaching law
N	Number of agents
n	number of states

List of Figures

Fig. 1.1 Quadcopter courtesy: Quanser Pvt. Ltd. 2
Fig. 1.2 Schematic diagram of Quadcopter 3
Fig. 1.3 Formation and tracking control for multi-agent system 4
Fig. 1.4 Literature review on formation protocol 6
Fig. 2.1 Multi-agent system representation 12
Fig. 2.2 Multi-sgent system in nature 13
Fig. 2.3 Graphical interpretation of SMC 17
Fig. 3.1 Quadcopter swarm. Courtesy: Quanser PVT. LTD 26
Fig. 3.2 Block diagram for formation of Quadcopter swarm 26
Fig. 3.3 Transitional motion of continuous-time primary control of Quadcopter 31
Fig. 3.4 Rotational motion of continuous-time primary control of Quadcopter 32
Fig. 3.5 Control efforts of continuous-time primary control of Quadcopter 33
Fig. 3.6 Trajectory tracking of continuous-time primary control of Quadcopter 33
Fig. 3.7 Communication topology 35
Fig. 3.8 X position of all agents using DCSM constant-proportional reaching law 40
Fig. 3.9 Y position of all agents using DCSM constant-proportional reaching law 40
Fig. 3.10 Z position of all agents using DCSM constant-proportional reaching law 41
Fig. 3.11 Tracking error of all agents using DCSM constant-proportional reaching law 41
Fig. 3.12 Control efforts of all agents using DCSM constant-proportional reaching law 42
Fig. 3.13 Sliding surface of all agents using DCSM constant-proportional reaching law 42
Fig. 3.14 X position of all agents using DCSOSM ST protocol 46

Fig. 3.15 Y position of all agents using DCSOSM ST protocol 46
Fig. 3.16 Z position of all agents using DCSOSM ST protocol 47
Fig. 3.17 Tracking error of all agents using DCSOSM ST protocol 47
Fig. 3.18 Control efforts of all agents using DCSOSM ST protocol 48
Fig. 3.19 Sliding surface of all agents using DCSOSM ST protocol 48
Fig. 3.20 Transitional motion of non-linear model of Quadcopter 55
Fig. 3.21 Control efforts of non-linear model of Quadcopter 55
Fig. 3.22 Sliding surface of non-linear model of Quadcopter 56
Fig. 3.23 Tracking error of non-linear model of Quadcopter 56
Fig. 3.24 Communication Topology 57
Fig. 3.25 X trajectory for all agents using DCSOSM ST protocol 57
Fig. 3.26 Y trajectory for all agents using DCSOSM ST protocol 59
Fig. 3.27 Z trajectory for all agents using DCSOSM ST protocol 60
Fig. 3.28 Control efforts for all agents using DCSOSM ST protocol 61
Fig. 3.29 Control efforts for all agents using DCSOSM ST protocol 61
Fig. 3.30 Sliding surface for all agents using DCSOSM ST protocol 62
Fig. 3.31 X & Y position for all agents under the disturbance 62
Fig. 4.1 Transitional motion of discrete-time primary control of Quadcopter 74
Fig. 4.2 Rotational motion of discrete-time primary control of Quadcopter 75
Fig. 4.3 Control effort of discrete-time primary control of Quadcopter 76
Fig. 4.4 Trajectory Tracking of discrete-time primary control of Quadcopter 77
Fig. 4.5 Communication Topology 85
Fig. 4.6 X position of all agents using DDSM power rate reaching law 86
Fig. 4.7 Y position of all agents using DDSM power rate reaching law 86
Fig. 4.8 Z position of all agents using DDSM power rate reaching law 87
Fig. 4.9 Tracking error of all agents using DDSM power rate reaching law 87
Fig. 4.10 Control efforts of all agents using DDSM power rate reaching law 88
Fig. 4.11 Sliding surface of all agents using DDSM power rate reaching law 88
Fig. 4.12 Results of all agents under disturbance 89
Fig. 4.13 X position of all agents using DDSM Gao's reaching law 95
Fig. 4.14 Y position of all agents using DDSM Gao's reaching law 96
Fig. 4.15 Z position of all agents using DDSM Gao's reaching law 96
Fig. 4.16 Tracking error of all agents using DDSM Gao's reaching law 97

Fig. 4.17 Control efforts of all agents using DDSM Gao's reaching law ... 97
Fig. 4.18 Sliding surface of all agents using DDSM Gao's reaching law ... 98
Fig. 5.1 Communication Topology ... 102
Fig. 5.2 Tracking of X position of all agents using DDSOSM ST protocol ... 107
Fig. 5.3 Tracking of Y position of all agents using DDSOSM ST protocol ... 108
Fig. 5.4 Tracking of Z position of all agents using DDSOSM ST protocol ... 108
Fig. 5.5 Tracking error of all agents using DDSOSM ST protocol ... 109
Fig. 5.6 Control effort of all agents using DDSOSM ST protocol ... 109
Fig. 5.7 Sliding surface of all agents using DDSOSM ST protocol ... 110
Fig. 5.8 X position of all agents using DDSM Bartoszewicz's reaching law ... 114
Fig. 5.9 Y position of all agents using DDSM Bartoszewicz's reaching law ... 115
Fig. 5.10 Z position of all agents using DDSM Bartoszewicz's reaching law ... 115
Fig. 5.11 Control effort of all agents using DDSM Bartoszewicz's reaching law ... 116
Fig. 5.12 Sliding surface of all agents using DDSM Bartoszewicz's reaching law ... 117
Fig. 5.13 Tracking error of all agents using DDSM Bartoszewicz's reaching law ... 117
Fig. 5.14 X position of all agents using DDSM Gao's reaching law ... 118
Fig. 5.15 Y position of all agents using DDSM Gao's reaching law ... 118
Fig. 5.16 Z position of all agents using DDSM Gao's reaching law ... 119
Fig. 5.17 Tracking error of all agents using DDSM Gao's reaching law ... 119
Fig. 5.18 Control effort of all agents using DDSM Gao's reaching law ... 120
Fig. 5.19 Sliding surface of all agents using DDSM Gao's reaching law ... 120
Fig. 6.1 Communication Topology ... 125
Fig. 6.2 X position of all agents using DDSM exponential reaching law ... 133
Fig. 6.3 Y position of all agents using DDSM exponential reaching law ... 133
Fig. 6.4 Z position of all agents using DDSM exponential reaching law ... 134
Fig. 6.5 Tracking error of all agents using DDSM exponential reaching law ... 134
Fig. 6.6 Control efforts of all agents using DDSM exponential reaching law ... 135

Fig. 6.7 Sliding surface of all agents using DDSM exponential reaching law 136
Fig. 6.8 Results of all agents under the disturbance 137
Fig. 6.9 X position of all agents using DDSM Gao's reaching law 138
Fig. 6.10 Y position of all agents using DDSM Gao's reaching law 138
Fig. 6.11 Z position of all agents using DDSM Gao's reaching law 139
Fig. 6.12 Tracking error of all agents using DDSM Gao's reaching law 139
Fig. 6.13 Control efforts of all agents using DDSM Gao's reaching law 140
Fig. 6.14 Sliding surface of all agents using DDSM Gao's reaching law 141

List of Tables

Table 3.1 System parameters of Quadcopter 28
Table 3.2 Comparison study of sliding variable ($\triangle_i$) 49
Table 3.3 Comparison study of control effort 49
Table 4.1 System parameters of Quadcopter 67
Table 4.2 Motor and Propellers specifications 67
Table 4.3 Specification of Battery 68
Table 4.4 Comparison study of QSMB ($\triangle_i$) 98
Table 4.5 Comparison study of control effort 98
Table 4.6 Number of steps required for the formation 99
Table 5.1 Comparison study of tracking error 121
Table 5.2 Comparison study of control effort 121
Table 6.1 Comparison study of QSMB ($\triangle_i$) 142
Table 6.2 Comparison study of control effort 142

Chapter 1
Introduction

1.1 Overview of Quadcopter Swarm

Recently, the interest in formation control of Unmanned Aerial Vehicles (UAVs) has significantly increased due to their several applications in defense, farming, rescuing, surveillance, and many more unexplored areas [1]. Broadly, there are two types of UAVs, namely, rotor-craft UAVs and Fixed wing UAVs. Compared to fixed-wing UAVs, rotor-craft UAVs have a couple of advantages, including the ability to hover in one place and perform vertical takeoff and landing (VTOL) [2]. Among all varieties of rotor-craft UAVs, the Quadcopter is the most popular due to its ability to hover and maneuver in tight spaces, making it ideal for tasks that require more precise positioning and control than traditional rotor-crafts. The Quadcopter features four BLDC motors coupled to four propellers given in Fig. 1.1. Each motor in a cross-sectional frame is positioned at the end of its arm. Also, the propeller rotation of the Quadcopter in a three-dimensional environment is made possible by the upward thrust forces. It features an onboard micro-controller that can stabilize the Quadcopter to the desired set point. It also has a battery connected to it, which may provide power and control to the motors through electronic speed controllers (ESCs).

Since a Quadcopter has four propellers and for each propeller speed and direction of the rotation are controlled by the controller to get the desired location. All four motors are placed at an equal distance from each other. To maintain the balance of the Quadcopter, one diagonal pair of the motors rotates in a clockwise direction and the other pair rotates in an anticlockwise direction [3]. Basically, a Quadcopter has two states to float in the air, where one is maintaining the desired location is called the hovering state and the second is continuously moving in the air is called the soaring state.

A. Mehta and A. Modi, *Robust Sliding Mode Protocols for Formation of Quadcopter Swarm*, Studies in Systems, Decision and Control 521,
https://doi.org/10.1007/978-981-99-9726-8_1

Fig. 1.1 Quadcopter courtesy: Quanser Pvt. Ltd.

Depending on these two states, motions can be classified into two types:

- Transitional Motion

  1. X-axis
  2. Y-axis
  3. Z-axis(altitude)

- Rotational Motion

  1. Roll
  2. Pitch
  3. Yaw

In this research work, the following motions are considered.

- Front and Backward motion = X-axis
- Left and Right motion = Y-axis
- Upwards and Downwards motion = Z-axis (Altitude)
- Rotation around the front-to-back axis = Roll
- Rotation around the side-to-side axis = Pitch
- Rotation around the vertical axis = Yaw.

From the free body diagram of the Quadcopter shown in Fig. 1.2, the motions are defined as follows: Forward motion will occur when the thrust generated by motors 2 and 3 is greater than motors 1 and 4. Similarly, Quadcopter will move backward if the thrust generated by motors 1 and 4 is greater than motors 2 and 3. If the thrust generated by motors 1 and 2 is greater than motors 3 and 4 then Quadcopter will move in the left direction and similarly the thrust generated by motors 3 and 4 is greater than motors 1 and 2 then the Quadcopter will move in the right direction. When all four motors generate an equally increasing thrust then the Quadcopter will move in the upward direction and vice versa for the downward motion. Roll motion will happen to rotate around the X-axis, pitch motion will occur to rotate around the Y-axis, and yaw motion will occur to rotate around the Z-axis.

A Quadcopter swarm, also known as a drone swarm, refers to a group of multiple Quadcopters working together in a coordinated manner to achieve a common goal. These swarm can be controlled autonomously through a central controller or through a distributed controller, allowing them to perform complex tasks that would be challenging or impossible for a single drone to accomplish. In general,

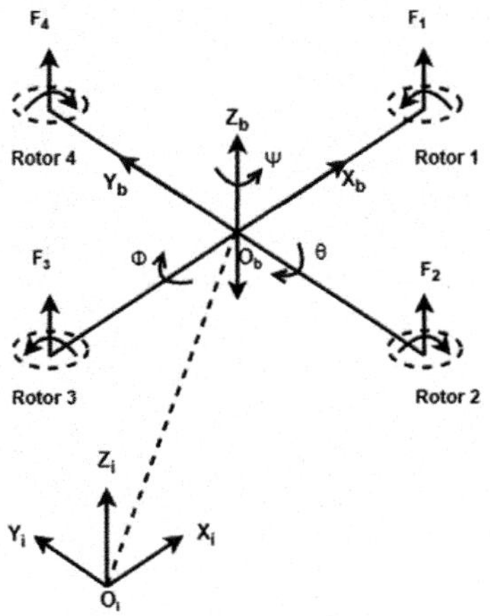

Fig. 1.2 Schematic diagram of Quadcopter

Quadcopters are frequently utilized aerial vehicles for formation flight due to their characteristics, and a wide range of civil applications. By leveraging the power of multiple Quadcopters working together, swarm technology offers several advantages over single drone operations, including increased efficiency, improved redundancy, and enhanced capabilities. Formation control is coordinated control where multiple Quadcopters follow a predefined trajectory. The goal of formation control is to maintain a specific spatial arrangement between the agents, while also ensuring that each Quadcopter avoids collisions with other Quadcopter and obstacles in the environment. Therefore it is believed that the formation of multiple low-cost Quadcopters is more effective in accomplishing the goal than a single high-tech Quadcopter [4]. In order to achieve the formation, the Multi-Agent System (MAS) concept has been used widely for controlling and synchronizing multiple systems like robotic arms, helicopter systems, and many more [5]. In each of these applications, the agents in the system work together to achieve a common goal, such as maximizing efficiency, minimizing congestion, or optimizing profits. MAS is a computational system made up of numerous autonomous agents also termed as systems that collaborate to accomplish shared or unique objectives [6]. MAS has grown in popularity due to its capability to address complicated problems that are challenging for single agents to manage.

As an illustration, Figure 1.3 shows the formation and tracking of MAS where two different types of agents, namely, unmanned aerial vehicles and autonomous ground vehicles have taken part. In this system, each individual system acts as an agent and all agents are communicating with their neighborhood agents to achieve the formation in a given pattern.

While using the application of MAS such as formation [7–9], flocking [10–12], rendezvous [13, 14], consensus [15, 16], etc. mainly there are two types of controllers,

- **Primary Controller:**
  - A primary controller is a device or software that regulates a process variable to maintain a desired setpoint.

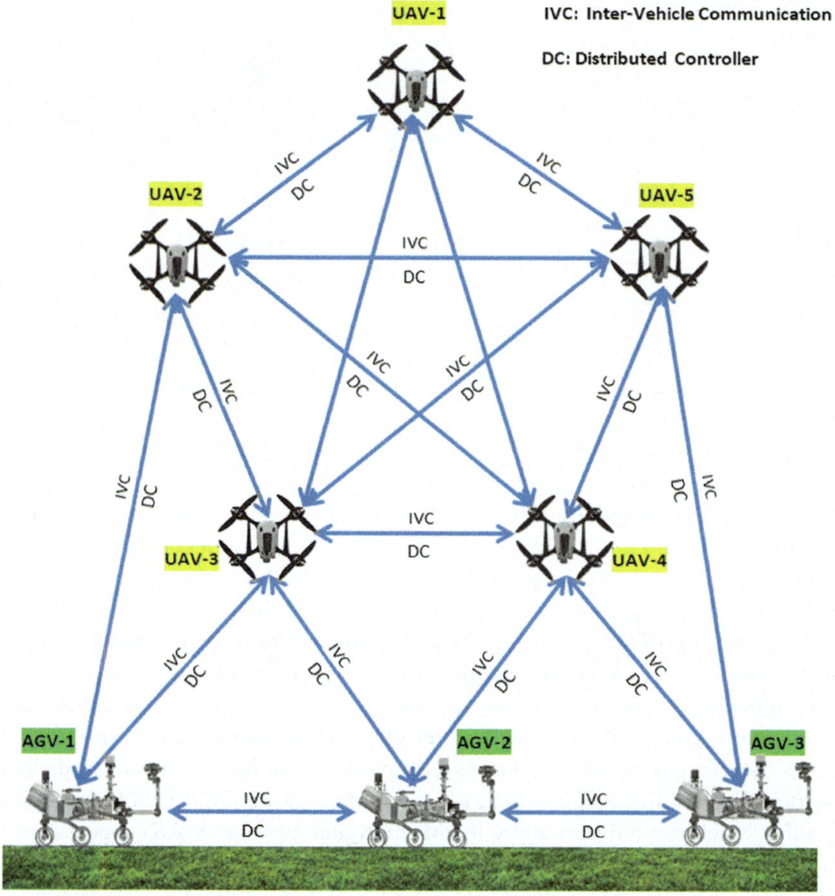

Fig. 1.3 Formation and tracking control for multi-agent system

- The primary controller receives input from sensors that measure the process variable and adjusts its output accordingly to maintain the desired setpoint.

- **Distributed or Secondary Controller:**
  - A distributed controller is mandated for the formation between all the Quadcopters. For this research work, distributed controller is referred as a formation control.
  - The formation controller receives input from the leader and adjusts the output for the desired setpoint and counts it to the primary controller.

Many researchers have contributed to the field of the formation of multiple Quadcopters. Few of them have designed the formation control of Quadcopter swarm using conventional controllers such as proportional-derivative (PD), proportional-integral-derivative (PID), model-predictive control (MPC), and linear quadratic regulator (LQR). However, conventional control techniques could not be adequately efficient to maintain the formation under any environmental uncertainty. So as a result, researchers have chosen the sliding mode control (SMC) for the advantages of robustness, finite-time convergence, and simplicity to design. In the context of Quadcopter swarms, sliding mode control offers several advantages.

Firstly, Quadcopter swarms inherently face uncertainties in terms of varying wind conditions, sensor noise, and communication delays. Sliding mode control's robustness against uncertainties enables the swarm to maintain formation and perform coordinated maneuvers even in the presence of such uncertainties. This robustness ensures the stability and reliability of the swarm in real-world scenarios. Secondly, sliding mode control is well-suited for handling the non-linear dynamics of Quadcopters. The complex aerodynamics, multi-variable interactions, and highly coupled dynamics of Quadcopters make them inherently non-linear systems. Sliding mode control can effectively handle these non-linearities without relying on precise mathematical models, making it an attractive choice for controlling Quadcopter swarms.

Moreover, sliding mode control can facilitate the formation control of Quadcopter swarms by ensuring that each Quadcopter converges to and maintains its desired position relative to its neighboring Quadcopters. The sliding mode control law generates control signals that guide each Quadcopter to follow a predefined formation pattern, allowing the swarm to achieve desired shapes, configurations, or trajectories. In the formation of Quadcopter swarms, the sliding surface is defined based on the desired formation pattern. The control law and switching strategy are designed to drive the Quadcopters toward the sliding surface and maintain them on it, ensuring that the desired formation is achieved and maintained. The sliding mode control approach allows for precise formation control, adaptability to changes in the formation, and robustness against disturbances.

1.2 Literature Review

The literature review on formation, as displayed in Fig. 1.4, covers a variety of research approaches. Numerous researchers [17–21, 28] across the world have provided their insightful perspectives on various MAS consensus approaches. There are several configurations of MAS which can generally be divided into two broad categories: Leaderless MAS and Leader-following MAS in which the agents synchronize with the commanded leader. Several different distributed control strategies such as formation, consensus, flocking, schooling, etc. have been proposed by several researchers [33, 37, 38]. The linear quadratic regulator (LQR) technique to develop the leader-following first-order and second-order consensus for MAS is utilized in [30]. Many authors [22–24] choose an event-triggered-based leader-following consensus in which control is updated only if the triggering condition is satisfied in order to conserve communication bandwidth and power resources. In addition, there are

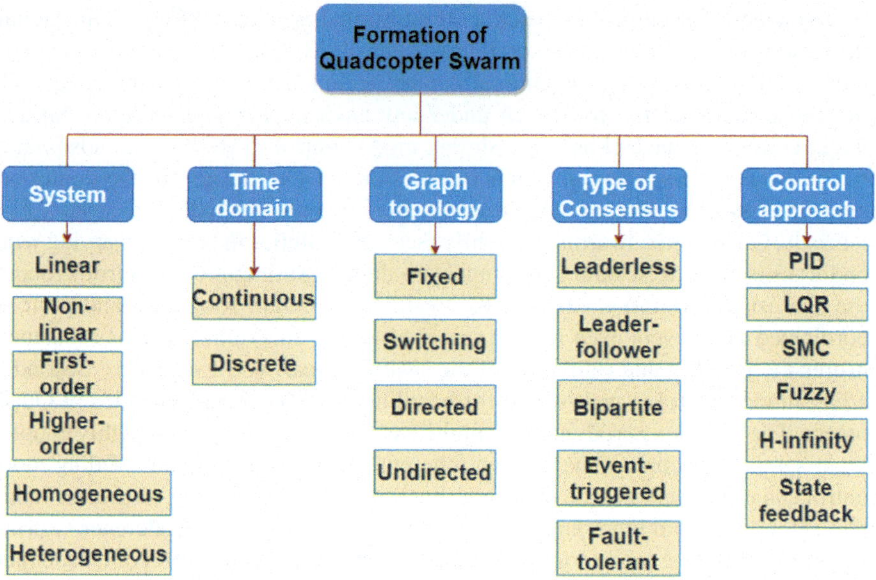

Fig. 1.4 Literature review on formation protocol

numerous more causes for MAS faults, including unavoidable system disturbances, sensor failure, actuator failure, and communication failure. As a result, numerous authors [25–27] have used the fault-tolerant approach to overcome this problem.

1.2.1 Technical Approaches for Formation of Quadcopter Swarm

Many researchers have contributed to the field of the formation of multiple Quadcopter swarm. The formation control of Quadcopters based on the leader-follower approach is proposed in [29]. A leader-follower consensus and formation control has been used for UAVs with event-triggered communication [31]. The application of multiple UAVs using sliding mode control where multi-UAV air threat assessment model has been described in [32]. The formation and tracking control for the Multi-agent network within fixed time convergence where formation moves in synchronization with the leader [34, 45]. In [35] the formation control with the external disturbances using sliding mode control is discussed. In [36], the leaderless formation protocol for the Quadrotors has been proposed using an undirected communication graph. The researchers in [35, 37] have proposed non-linear Quadcopter model-based formation control using sliding mode control.

Several formation control designs of multiple Quadcopters have been proposed in literature such as adaptive sliding mode control [39], model-predictive control [40], and event-triggered integral sliding mode control [41] with unpredictable disturbances. In [44], the formation tracking with a dynamic leader for time-varying multi-UAV system is proposed.

In the above formation control methods, the homogeneous UAV system has been considered. Simultaneously researchers have worked on heterogeneous MAS for leader-follower consensus using MPC [52], LQR control [51], and integral sliding mode control (ISMC) [46]. Recently, [49] has proposed a super-twisting algorithm for the formation control for the Quadrotors under the external disturbances and directed topology in the continuous-time domain. Moreover, an event-triggered controller is proposed for the formation of a time-delayed discrete-time MAS of multi-UAV in [43] and in [48] time-varying formation control second-order DMAS with switching topologies are investigated.

1.3 Organization of Book

The structure of the book is as follows:

- **Chapter** 1 provides a brief introduction to the formation of Quadcopter swarm and literature survey on the different formation control algorithms for Quadcopter swarm in continuous and discrete-time domain.
- **Chapter** 2 gives the detailed information for the sliding mode control, overview of the MAS concept, and preliminaries of graph theory for developing the protocols.
- **Chapter** 3 provides the detailed information about the design of sliding mode protocols for the formation of a linear and non-linear Quadcopter swarm in continuous-time.
- **Chapter** 4 presents the design of sliding mode protocols for the formation of a Quadcopter swarm along with the number of steps required for the formation and Lyapunov stability analysis in discrete-time.
- **Chapter** 5 presents the design of discrete-time higher order sliding mode (DHSM) protocol for the formation of a Quadcopter swarm.
- **Chapter** 6 presents the design of discrete-time sliding mode protocol using exponential reaching law for the formation of a Quadcopter swarm along the number of steps required for the formation.
- **Chapter** 7 finishes the book with a summary of the research's main findings. Future research possibilities are also discussed.

References

1. Shakhatreh, H., et al.: Unmanned aerial vehicles (UAVs): a survey on civil applications and key research challenges. IEEE Access **7**, 48572–48634 (2019). https://doi.org/10.1109/ACCESS.2019.2909530
2. Stöcker, C., Bennett, R., Nex, F., Gerke, M., Zevenbergen, J.: Review of the current state of UAV regulations. Remote. Sens. **9**(5), 459 (2017). https://doi.org/10.3390/rs9050459
3. Kim, S.J., Jeong, Y., Park, S., Ryu, K., Oh, G.: A survey of drone use for entertainment and AVR (augmented and virtual reality). In: Jung, T., Tom Dieck, M. (eds) Augmented Reality and Virtual Reality. Progress in IS. Springer, Cham (2018). https://doi.org/10.1007/978-3-319-64027-3_23
4. Mahmood, A., Kim, Y.: Leader-following formation control of quadcopters with heading synchronization. Aerosp. Sci. Technol. **47**, 68–74 (2015). ISSN 1270-9638. https://doi.org/10.1016/j.ast.2015.09.009
5. Patel, K., Mehta, A.: Discrete-Time Sliding Mode Protocols for Discrete Multi-Agent System, vol. 303, pp. 1–121. Springer (2021). https://doi.org/10.1007/978-981-15-6311-9
6. Fang, L.: Cooperative control of multi-agent systems with information flow constraints. Ph.D. thesis, University of Notre Dame (2006)
7. Fax, J.A., Murray, R.M.: Information flow and cooperative control of vehicle formations. IEEE Trans. Autom. Control. **49**, 1465–1476 (2004)
8. Ren, W.: Consensus strategies for cooperative control of vehicle formations. IET Control. Theory Appl. **1**, 505–512 (2007)
9. Tomic, I., Milonidis, E., Halikias, G.D.: LQR distributed cooperative control of a formation of low-speed experimental UAVs. In: 2016 UKACC 11th International Conference on Control (CONTROL), pp. 1–6 (2016)
10. Chuang, Y., Huang, Y.R., D'Orsogna, M.R., Bertozzi, A.L.: Multi-vehicle flocking: scalability of cooperative control algorithms using pairwise potentials. In: Proceedings 2007 IEEE International Conference on Robotics and Automation, pp. 2292–2299 (2007)
11. La, H.M., Sheng, W.: Flocking control algorithms for multiple agents in cluttered and noisy environments. In: Bio-Inspired Self-Organizing Robotic Systems, pp. 53–79. Springer, Berlin (2011)
12. Zhao, W., Chu, H., Zhang, M., Sun, T., Guo, L.: Flocking control of fixed-wing UAVs with cooperative obstacle avoidance capability. IEEE Access **7**, 17798–17808 (2019)
13. Dimarogonas, D.V., Kyriakopoulos, K.J.: On the rendezvous problem for multiple nonholonomic agents. IEEE Trans. Autom. Control. **52**, 916–922 (2007)
14. Lin, P., Ren, W., Wang, H., Al-Saggaf, U.M.: Multiagent rendezvous with shortest distance to convex regions with empty intersection: algorithms and experiments. IEEE Trans. Cybern. **49**, 1026–1034 (2019)
15. He, D., Shi, D., Sharma, R.: Consensus-based distributed cooperative control for microgrid voltage regulation and reactive power sharing. In: IEEE PES Innovative Smart Grid Technologies, Europe, pp. 1–6 (2014)
16. Islam, S., Liu, P.X., Saddik, A.E.: Consensus based distributed cooperative control for multiple miniature aerial vehicles with uncertainty. In: 2017 IEEE International Conference on Systems, Man, and Cybernetics (SMC), pp. 2477–2479 (2017)
17. Tao, L., Jifeng, Z.: Mean square average consensus of multi-agent systems with time-varying topologies and stochastic communication noises. In: 2008 27th Chinese Control Conference, pp. 552–556 (2008)
18. Wang, F., Yang, H.: Containment consensus of multi-agent systems with communication noises. In: Proceedings of 2016 Chinese Intelligent Systems Conference, pp. 189–197. Springer, Singapore (2016)
19. Parsegov, S., Polyakov, A., Shcherbakov, P.: Fixed-time consensus algorithm for multi-agent systems with integrator dynamics. IFAC Proc. Vol. **46**(27), 110–115 (2013). In: 4th IFAC Workshop on Distributed Estimation and Control in Networked Systems (2013)

20. Chen, S., Ho, D.W.C., Li, L., Liu, M.: Fault-tolerant consensus of multi-agent system with distributed adaptive protocol. IEEE Trans. Cybern. **45**, 2142–2155 (2015)
21. Mu, X., Xiao, X., Liu, K., Zhang, J.: Leader-following consensus of multi-agent systems with jointly connected topology using distributed adaptive protocols. J. Frankl. Inst. **351**(12), 5399–5410 (2014)
22. Li, H., Liao, X., Huang, T., Zhu, W.: Event-triggering sampling based leader-following consensus in second-order multi-agent systems. IEEE Trans. Autom. Control. **60**, 1998–2003 (2015)
23. Xu, B., He, W., Ye, D.: Event-triggered consensus for general linear leader-following multiagent systems under directed topologies. In: IECON 2018-44th Annual Conference of the IEEE Industrial Electronics Society, pp. 5971–5976 (2018)
24. Tan, X., Cao, J., Li, X.: Consensus of leader-following multiagent systems: a distributed event-triggered impulsive control strategy. IEEE Trans. Cybern. **49**, 792–801 (2019)
25. Yang, P., Xu, M., Li, D., Liu, Z., Huang, Y.: Distributed fault tolerant consensus control for multi-agent system with actuator fault based on adaptive observer. Trans. Inst. Meas. Control. **41**, 4207–4217 (2019)
26. Fattahi, M., Afshar, A.: Fault tolerant consensus control of multi-agent systems with switching topology. SN Appl. Sci. **1**, (2019)
27. Chen, S., Ho, D.W.C., Li, L., Liu, M.: Fault-tolerant consensus of multi-agent system with distributed adaptive protocol. IEEE Trans. Cybern. **45**, 2142–2155 (2015)
28. Zhu, W., Cheng, D.: Leader-following consensus of second-order agents with multiple time varying delays. Automatica **46**(12), 1994–1999 (2010)
29. Xuan-Mung, N., Hong, S.K.: Robust adaptive formation control of quadcopters based on a leader-follower approach. Int. J. Adv. Robot. Syst. **16**(4), (2019). https://doi.org/10.1177/1729881419862733
30. Ma, J., Zheng, Y., Wang, L.: LQR-based optimal topology of leader-following consensus. Int. J. Robust Nonlinear Control. **25**(17), 3404–3421
31. Guerrero-Castellanos, J.F., Vega-Alonzo, A., Durand, S., Marchand, N., Gonzalez-Diaz, V.R., Castañeda-Camacho, J., Guerrero-Sánchez, W.F.: Leader-following consensus and formation control of VTOL-UAVs with event-triggered communications. Sensors **19**(24), 5498 (2019). https://doi.org/10.3390/s19245498
32. Liu, C., Sun, S., Tao, C., Shou, Y., Xu, B.: Sliding mode control of multi-agent system with application to UAV air combat. Comput. Electr. Eng. https://doi.org/10.1016/j.compeleceng.2021.107491
33. Wen, G., Hu, G., Yu, W., Cao, J., Chen, G.: Consensus tracking for higher-order multi-agent systems with switching directed topologies and occasionally missing control inputs. Syst. Control. Lett. **62**(12), 1151–1158 (2013). ISSN 0167-6911. https://doi.org/10.1016/j.sysconle.2013.09.009
34. Xu, G.-H., Li, M., Chen, J., Lai, Q., Zhao, X.-W.: Formation tracking control for multi-agent networks with fixed time convergence via terminal sliding mode control approach. Sensors **21**(4), 1416 (2021). https://doi.org/10.3390/s21041416
35. Nguyen, N.P., Park, D., Ngoc, D.N., Xuan-Mung, N., Huynh, T.T., Nguyen, T.N., Hong, S.K.: Quadrotor formation control via terminal sliding mode approach: theory and experiment results. Drones **6**(7), 172 (2022). https://doi.org/10.3390/drones6070172
36. Mazhar, N., Malik, F. M., Khan, R., Raza, A., Mazhar, S., Irfan, M.: Robust decentralized nonlinear formation control of multiagent quadrotor. In: 2020 7th International Conference on Electrical and Electronics Engineering (ICEEE), pp. 303–309. IEEE (2020)
37. Patel, K., Mehta, A.: Fixed time steps discrete-time sliding mode consensus protocols for two degree of freedom helicopter systems. Trans. Inst. Meas. Control. **43**(15), 3322–3334 (2021). https://doi.org/10.1177/01423312211004778
38. Patel, K., Mehta, A.: Discrete-time event-triggered higher order sliding mode control for consensus of 2-DOF robotic arms. Eur. J. Control. **56**, 231–241 (2020). ISSN 0947-3580. https://doi.org/10.1016/j.ejcon.2020.01.006
39. Wang, Z., Fei, Q., Wang, B.: Distributed adaptive sliding mode formation control for multiple unmanned aerial vehicles. In: Chinese Control And Decision Conference (CCDC), Hefei, China, vol. 2020, pp. 2105–2110 (2020). https://doi.org/10.1109/CCDC49329.2020.9164781

40. Vargas, S., Becerra, H.M., Hayet, J.-B.: MPC-based distributed formation control of multiple quadcopters with obstacle avoidance and connectivity maintenance. Control. Eng. Pract. **121**, 105054 (2022). ISSN 0967-0661. https://doi.org/10.1016/j.conengprac.2021.105054
41. Hou, Z., Lu, P.: Event-triggered integral sliding mode formation control for multiple quadrotor UAVs with unknown disturbances. Frankl. Open **1**, 17–29 (2022). ISSN 2773-1863. https://doi.org/10.1016/j.fraope.2022.05.004
42. González-Sierra, J., Dzul, A., Martínez, E.: Formation control of distance and orientation based-model of an omnidirectional robot and a quadrotor UAV. Robot. Auton. Syst. **147**, 103921 (2022). ISSN 0921-8890. https://doi.org/10.1016/j.robot.2021.103921
43. Yan, Z., Han, L., Li, X., Dong, X., Li, Q., Ren, Z.: Event-triggered formation control for time-delayed discrete-time multi-agent system applied to multi-UAV formation flying. J. Frankl. Inst. **360**(5), 3677–3699 (2023). ISSN 0016-0032. https://doi.org/10.1016/j.jfranklin.2023.01.036
44. Wang, J., Han, L., Dong, X., Li, Q., Ren, Z.: Distributed sliding mode control for time-varying formation tracking of multi-UAV system with a dynamic leader. Aerosp. Sci. Technol. **111**, 106549 (2021). ISSN 1270-9638. https://doi.org/10.1016/j.ast.2021.106549
45. Song, R., Xing, S., Xu, Z.: Finite-time leader-follower consensus of a discrete-time system via sliding mode control. Front. Inf. Technol. Electron. Eng. **23**(7), 1057–1068 (2022). https://doi.org/10.1631/FITEE.2100565
46. Mu, B., Zhang, K., Shi, Y.: Integral sliding mode flight controller design for a quadrotor and the application in a heterogeneous multi-agent system. IEEE Trans. Ind. Electron. **64**(12), 9389–9398 (2017). https://doi.org/10.1109/TIE.2017.2711575
47. De Queiroz, M., Cai, X., Feemster, M.: Formation Control of Multi-Agent Systems: a Graph Rigidity Approach. John Wiley and Sons (2019). https://doi.org/10.1002/9781118887455
48. He, L., Zhang, J., Hou, Y., Liang, X., Bai, P.: Time-varying formation control for second-order discrete-time multi- agent systems with switching topologies and nonuniform communication delays. IEEE Access **7**, 65379–65389 (2019). https://doi.org/10.1109/ACCESS.2019.2918864
49. Shao, S., Zhao, Y., Wu, X.: Robust fixed-time formation control for quadrotor unmanned aerial vehicles with directed topology. Trans. Inst. Meas. Control. **44**(2), 324–338 (2022). https://doi.org/10.1177/01423312211032541
50. Derakhshannia, M., Moosapour, S.S.: Disturbance observer-based sliding mode control for consensus tracking of chaotic nonlinear multi-agent systems. Math. Comput. Simul. **194**, 610–628 (2022). ISSN 0378-4754. https://doi.org/10.1016/j.matcom.2021.12.017
51. Mu, B., Shi, Y.: Distributed LQR consensus control for heterogeneous multiagent systems: theory and experiments. IEEE/ASME Trans. Mechatron **23**(1), 434–443 (2018). https://doi.org/10.1109/TMECH.2018.2791544
52. Gopakumar, K., Shihabudheen, K.V.: Leader Follower Distributed Consensus Control of Heterogeneous Multi-agent System with Model Predictive Control. In: International Symposium of Asian Control Association on Intelligent Robotics and Industrial Automation (IRIA), Goa, India, vol. 2021, pp. 25–30 (2021). https://doi.org/10.1109/IRIA53009.2021.9588671

Chapter 2
Preliminaries of Multi-agent System, Graph Theory, and Sliding Mode Control

2.1 Overview of Multi-agent System

A branch of artificial intelligence known as distributed artificial intelligence (DAI) has grown significantly in relevance as a result of its capacity to address challenging real-world problems [1]. The three major areas that have received the most attention in this field of study artificially distributed intelligence. These include parallel AI, Multi-agent systems (MAS), and distributed problem-solving (DPS). The MAS is a computational framework that consists of multiple autonomous entities, known as agents, that have the capability to interact with their environment and each other. Each agent possesses its own set of objectives, information, and decision-making capabilities. Unlike traditional single-agent systems, MAS focuses on the collective behavior and coordination of multiple agents working together toward a common goal. These agents can perceive their surroundings, process information, make independent decisions, and execute actions based on their internal knowledge and local observations. The interactions among agents can lead to emergent behavior where the overall system behavior arises from the interactions of individual agents. MAS is characterized by its decentralized nature, allowing agents to operate independently without the need for a central controlling authority. This autonomy enables adaptability to dynamic environments and the ability to handle complex tasks that may be beyond the capabilities of a single agent. One of the key strengths of MAS lies in its versatility and applicability to a wide range of domains. It finds applications in fields such as robotics, artificial intelligence, economics, traffic management, energy systems, and more. Through the coordination, communication, and collaboration of agents, MAS provides a powerful framework for solving complex problems and achieving collective objectives in a decentralized and adaptive manner.

In contrast to distributed problem-solving, MAS focuses on the behavior of the computational entities that can be used to solve a given problem. To solve issues that are too complicated for a single agent to manage alone, MAS referred to as a network of individual agents can share information and interact with one another.

A. Mehta and A. Modi, *Robust Sliding Mode Protocols for Formation of Quadcopter Swarm*, Studies in Systems, Decision and Control 521,
https://doi.org/10.1007/978-981-99-9726-8_2

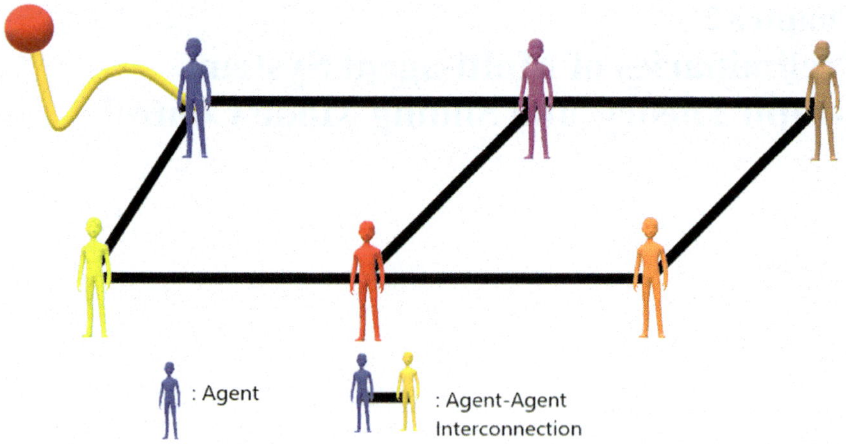

Fig. 2.1 Multi-agent system representation

Understanding the characteristics of each individual agent or computer unit is crucial for differentiating between a simple distributed system and a MAS. A MAS is a development of agent technology in which a collection of loosely coupled autonomous agents act in a setting to accomplish a shared objective. This is accomplished through either collaboration or competition, knowledge sharing or knowledge non-sharing, etc. The MAS has gained much attention in recent applications due to offering too many advantages for performing a particular task [2]. The adavantages of the MAS are given below:

- An increase in the efficiency and speed of the operation as the parallel computation has been done with asynchronous operation ability.
- An increment in the robustness of the system due to the capability of the system for taking over the situation in the condition of failure of any agent.
- More flexibility and scalability due to the capability of adding or removing the agent in the system.
- Re-usability of the system, as agents can easily be used or replaced in other systems or structures.

The MAS as depicted in Fig. 2.1, is made up of a number of interdependent intelligent agents that can be homogeneous, heterogeneous, or autonomous in nature. These agents interact with one another and communicate in order to solve complex problems that are too difficult or impossible for any one agent to handle alone. A simple system dynamics is illustrated by the word "agent", which can refer to a drone, a wheeled mobile robot, a helicopter system, etc.

The coordination of the MAS is the fundamental worry for playing out any undertaking as all specialists perform exclusively for achieving the specific objective. Thus, there is a need to synchronize all the specialists for keeping up with the strength of

(a) Flocking (b) Formation

(c) Fish Schooling (d) Ant Swarm

Fig. 2.2 Multi-sgent system in nature

the framework. This coordination is utilized in quite a large number of applications like satellite synchronous flying, robotized expressway frameworks, aviation authority, different versatile robot frameworks, and so on. Another fundamental concern is about the agreement which is characterized as the circulated control strategies for getting the shared objective of the specialists attempt to achieve [3]. The agreeable control of MAS utilized different applications like independent submerged vehicles, automated ethereal vehicles, particularly for military use, versatile robots, sustainable power, and a lot more neglected regions. Consistently, a typical concentrated advanced PC was utilized to gather the data from the other framework/specialist for dealing with the multi-specialist framework. The different types the cooperative behavior of MAS available in nature are shown in Fig. 2.2.

A MAS can be divided into two sorts based on the internal architecture of the specific individual agents [4]:

1. Homogeneous Structure:
   The homogeneous structure stands for the group of Quadcopters having the same characteristics, architecture, dynamics, and working throughout the operation. The homogeneous structure of a Quadcopter swarm revolves around the idea that all agents within the system share identical or very similar attributes, capabilities,

and behaviors. This allows them to work together efficiently and effectively to achieve common objectives. The agents can be distinguished from one another based on their physical position and the region of their surroundings where the activity is carried out.

2. Heterogeneous Structure:
   A heterogeneous architecture allows for a range in the agents' abilities, makeup, and functionality. Depending on the dynamics of the environment and the agent's unique location, the agent's behaviors may differ from those of an agent placed in a different area but they will still serve the same goal. Heterogeneous architecture significantly increases the relevance of modeling applications. It's possible for each agent to have particular local goals that clash with those of other agents. The video game Predator-Prey serves as an illustration of this. Both the predator and the prey can be modeled as agents in this scenario. The objectives of the two agents are almost definitely completely at odds with one another.

2.2 Preliminaries of Algebraic Graph Theory

Graph theory has evolved in the domain of control for representing information exchange among the agents in a MAS and to define the arithmetic flow and data association in computer networks [5]. Essentially, the graph theory has been used to model the relationships among agents of the MAS where the vertices of the graph correspond to the agents and the edges model the relation among these agents. There are two types of graphs:

1. Directed graph
   A graph is referred to as being directed if it contains an ordered pair of vertexes. A specific direction from one vertex to another vertex is defined by the graph's edges.
2. Undirected graph
   There is no specific direction to define the edges in an undirected graph, which is a graph with an unordered pair of vertexes.

A graph $\mathbb{G}$ with vertices $\mathbb{V}$ and edges $\mathcal{E}$ is commonly represented as $\mathbb{G} = (\mathbb{V}, \mathcal{E})$ [6, 7]. The number of vertices (nodes) and edges will be denoted by $\mathbb{V}$ and $\mathcal{E}$. The degree of a graph means more edges are there in vertices.

In the current state of graph theory, we define some matrices. $\mathbb{D}$ is a diagonal matrix that contains information about the number of edges on each vertex. The number of neighbors of a vertex determines its degree. The matrix $\mathbb{D}[\mathrm{i}, \mathrm{j}]$ where i and j represent the vertices of a graph is represented as

$$\mathbb{D}[\mathrm{i}, \mathrm{j}] = \begin{cases} \deg(\mathbb{V}_i) & \text{if; i = j} \\ 0 & \text{otherwise} \end{cases} \tag{2.1}$$

Here, deg($\mathbb{V}_i$) stands for the vertex i's degree.

A square matrix called the adjacency matrix $\mathbb{A}$ provides details regarding the proximity of vertices in an undirected graph. An adjacency matrix is defined as

$$\mathbb{A}[i, j] = \begin{cases} 1 \text{ if an edge exists between vertex } i \text{ and } j \\ 0 \text{ if an edge does not exist between vertex } i \text{ and } j \end{cases} \tag{2.2}$$

Using the concepts of degree matrix $\mathbb{D}$ and adjacency matrix $\mathbb{A}$, the Laplacian matrix can be defined as follows:

$$\mathbb{L} = \mathbb{D} - \mathbb{A} \tag{2.3}$$

The Laplacian matrix for an undirected graph can also be defined directly as follows:

$$\mathbb{L}[i, j] = \begin{cases} \deg(\mathbb{V}_i) & \text{if; vertex } i = j \\ 1 & \text{if; } i \neq j \text{ are adjacent} \\ 0 & \text{otherwise} \end{cases} \tag{2.4}$$

The Laplacian matrix is a symmetric and positive semi-definite matrix in special things. Its lowest eigenvalue is always zero and all other eigenvalues are positive and real. Laplacian matrix is diagonal in undirected topologies.

The pinning node is regarded as the "leader" of the pinning control method. The external signal controls only agents that communicate with the pinning node for pinning control. Pinning control methods can achieve agreement on non-identical target values.

We define the set of pinning nodes as follows: $\mathbb{P} = \text{diag}\{P_1, P_2 \ldots P_n\}$ such that,

$$\mathbb{P}_i = \begin{cases} 1 & \text{if node i is pinned} \\ 0 & \text{otherwise} \end{cases} \tag{2.5}$$

In the cooperative action of MAS, the pining node is called a leader, and other existing nodes are called followers of that leader. In a MAS, agents are only local views, goals, and information shared with other agents, and they can interact with other agents and work together to get the desired overall system behavior.

2.3 Review of Sliding Mode Control

The globe was initially introduced to a variable structure system with Sliding Mode Control (SMC) in the 1960s in the Soviet Union by [8]. Sliding mode control is a robust control technique that has gained significant attention in recent years due to its ability to handle uncertainties, disturbances, and non-linear dynamics in control systems. It offers an effective solution for controlling complex systems where precise mathematical models or extensive calculations may be challenging or unavailable

[9]. Sliding mode control ensures that the system's states converge and remain on a predefined sliding surface, providing robustness against uncertainties and disturbances.

The fundamental principles of sliding mode control lie in the concept of a sliding surface and the control objective. The sliding surface is a hyperplane defined in the state space, representing a desired condition or goal for the system. SMC involves creating a control law that guides the system's state trajectory along a specific mathematical surface called the sliding surface. The objective is to force the system's state to remain on this sliding surface, achieving the desired control objective. The control law operates in two modes: sliding mode and reaching mode. In the sliding mode, the control law drives the system along the sliding surface. In the reaching mode, the control law guides the system toward the sliding surface. The control objective is to drive the system's states to converge onto this sliding surface and maintain them on it. The control law in sliding mode control consists of two main components: a sliding surface function and a control function. The discontinuous surfaces in the phase planes are a result of the states at which the structures change. These surfaces are known as the sliding surfaces [10].

The sliding surface function determines the desired behavior and properties of the sliding surface, while the control function generates the control signal that drives the system toward the sliding surface. The control function often uses a discontinuous or switching strategy, which switches between different control actions based on the system's state and the proximity to the sliding surface. Sliding mode control is also well-suited for non-linear systems. It can handle systems with non-linear dynamics without requiring precise mathematical models or extensive calculations. The control action depends mainly on the system's states and the proximity to the sliding surface, making it suitable for controlling complex and non-linear systems.

The design of sliding mode control involves selecting an appropriate sliding surface and designing the control law. The choice of the sliding surface depends on the control objectives and desired system behavior. Different sliding surfaces can be used to achieve specific control objectives, such as tracking a reference trajectory or stabilizing the system around a desired equilibrium point. The control law design involves designing the control function and determining the switching strategy to generate the control signal. However, sliding mode control is also associated with a characteristic phenomenon called "chattering" [11]. Chattering refers to the high-frequency oscillations that can occur around the sliding surface due to the discontinuous control action. While chattering does not affect the system's stability, it can introduce high-frequency noise and can be problematic in certain applications, especially those involving sensitive devices or systems [12]. Various techniques have been proposed to mitigate chattering, such as introducing a boundary layer around the sliding surface or employing adaptive control strategies.

2.3.1 Continuous-Time Sliding Mode Control

In the sliding mode control procedure, a discontinuous action control consists of the following two phases [13, 14] as shown in Fig. 2.3:

1. Reaching phase
   In Reaching phase the state trajectories of the system move along toward the switching surface from their own initial conditions. In the reaching phase specific control effort is provided for the state trajectory reach to the defined sliding surface.
2. Sliding phase
   Sliding phase shows the sliding motion of trajectories of the system on the switching surface after reaching the sliding surface. A discontinuous control action is applied to structures for the state trajectory to stay on the switching surface. By that structures move toward the equilibrium point on the phase plane and gain robust control after reaching which shows the asymptotic stability of the system.

According to Utkin, a successful sliding mode in phase plane is possible if it will obey a sufficient condition which is described as

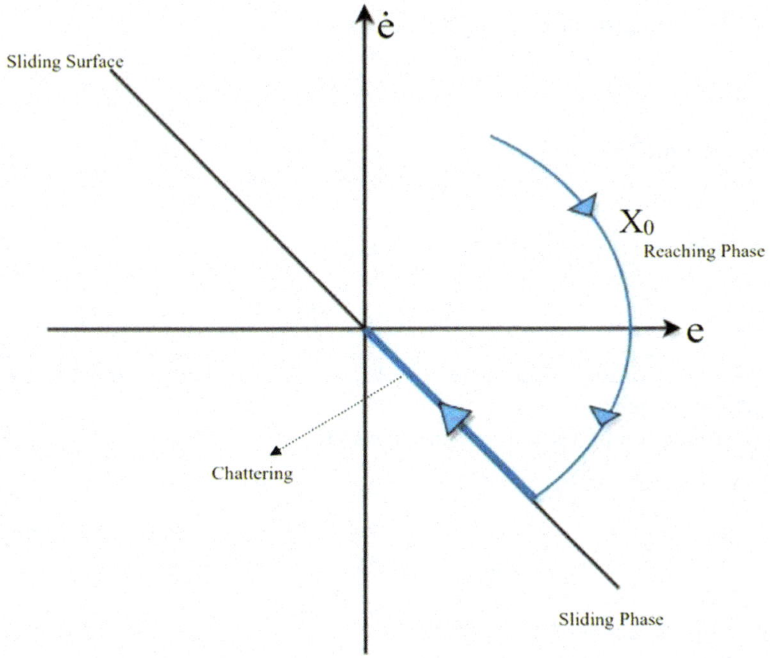

Fig. 2.3 Graphical interpretation of SMC

$$\lim_{s \to -0} \dot{s} > 0 \quad \text{and} \quad \lim_{s \to +0} \dot{s} < 0 \tag{2.6}$$

where s is defined as the sliding surface and $\dot{s}$ is defined as the rate of change of distance from the sliding surface. Now, a reachability condition is described as a condition on the system trajectories to reach the sliding surface for minimum finite convergence into the sliding motion [9].

$$s\dot{s} < 0 \tag{2.7}$$

So, for getting asymptotic stability in the system of sliding mode control, the reachability condition is necessary to satisfy. Generally, variable structure control has non-deteriorating oscillations/chattering on a sliding surface with a finite though very large frequency and range.

As previously mentioned, there are three components that make up the design process for sliding mode control: (i) reaching law (ii) sliding surface, and (iii) equivalent switching control. Let's consider a linear time-invariant (LTI) system as a basis for understanding the full sliding mode controller design process.

$$\dot{X}(t) = Ax(t) + Bu(t)$$

$$Y(t) = Cx(t). \tag{2.8}$$

Now, the sliding surface can be defined as

$$\mathbb{S} = \Upsilon X(t) \tag{2.9}$$

differentiate Eq. (2.9)

$$\dot{\mathbb{S}} = \Upsilon \dot{X}(t) \tag{2.10}$$

substituting the $\dot{X}(t)$ into Eq. (2.10), we may get

$$\dot{\mathbb{S}} = \Upsilon(Ax(t) + Bu(t)) \tag{2.11}$$

where Υ is the sliding mode gain which can be evaluated by the LQR or Pole Placement approach.

The introduction of several reaching laws [15, 16]:

- Constant reaching law

$$\dot{\mathbb{S}} = -\Lambda_1 sgn(\mathbb{S}), \ \Lambda_1 > 0. \tag{2.12}$$

- Constant-proportional rate reaching law

$$\dot{\mathbb{S}} = -\Lambda_1 \mathbb{S} - \Lambda_2 sgn(\mathbb{S}), \ \Lambda_1 \ \& \ \Lambda_2 > 0. \tag{2.13}$$

- Power rate reaching law

$$\dot{\mathbb{S}} = -\Lambda_1 \mathbb{S} - \Lambda_2 |\mathbb{S}|^{\eta} sgn(\mathbb{S}),\ 0 < \eta < 1. \tag{2.14}$$

- Power rate exponential reaching law

$$\dot{\mathbb{S}} = -\frac{\Lambda_1}{\mathcal{N}} sgn(\mathbb{S}),\ \mathcal{N} = \Phi + (1 - \Phi) e^{-\mathcal{F}|\mathbb{S}|^{\mathcal{P}}}. \tag{2.15}$$

Hence, the control law is derived using (2.8) and (2.9) and can be written as

$$u(t) = -(\Upsilon B)^{-1} [A(t)\Upsilon x(t) + \Lambda_1 sgn(\mathbb{S})]. \tag{2.16}$$

The fundamental disadvantage of CSMC is that the discontinuous control signal switches at a very high frequency to generate the sliding motion, which is known as chattering in literature, once the closed-loop system trajectories reach the sliding surface [17, 18]. Chattering in a closed loop is dangerous because it can stimulate unmodeled dynamics and cause the actuator to undergo excessive wear and tear, especially in mechanical systems. Due to the sample frequency's restriction, SMC allows low switching frequencies in discrete time, making it more practical for implementation. In order to reduce the chattering effect, the higher order CSMC is being used.

Continuous-time Higher order Sliding Mode Control

Higher Order Sliding Mode Control (HOSMC) is an advanced control strategy used in engineering and control systems. It builds upon the principles of sliding mode control, a robust technique for handling uncertainties and disturbances in a system. In traditional sliding mode control, a system's states are manipulated to slide along a specific manifold, or hyperplane, ensuring robustness and stability. HOSMC takes this a step further by incorporating higher order derivatives of the sliding surface into the control law. This means that not only the first derivative (as in traditional SMC), but also higher derivatives are utilized to stabilize the system.

HOSMC offers several advantages. It often leads to improved tracking performance, especially in systems with complex dynamics or non-linearities. Additionally, it tends to reduce the chattering phenomenon, where control signals switch rapidly. This results in smoother control actions. HOSMC is particularly effective in handling systems with high-order dynamics that cannot be effectively controlled using first-order sliding mode control.

- **Continuous-time Super-twisting Algorithm**

Super-twisting algorithm (STA) is a second-order sliding mode (SOSM) control method that aims to improve robustness and reduce chattering effects. STA is designed for the system having the relative degree-1 with respect to sliding mode variable. The switching law of super-twisting algorithm u_{sw} is given as [19]

$$u_{sw}(t) = -\Lambda_1\sqrt{|\mathbb{S}(t)|}sgn(\mathbb{S}(t)) + \omega(t)$$

$$\dot{\omega}(t) = -\Lambda_2 sgn(\mathbb{S}(t)). \tag{2.17}$$

where $\Lambda_1, \Lambda_2 > 0$.

The ultimate control law of super-twisting algorithm u(t) is a combination of two terms,

$$u(t) = u_{sw}(t) + u_{eq}(t) \tag{2.18}$$

where, $u_{eq}(t)$ is a equivalent control law. For finding $u_{eq}(t)$ let's consider $\dot{\mathbb{S}}(t) = 0$ in Eq. (2.11) and derive as

$$u_{eq}(t) = -(\Upsilon B)^{-1}[A\Upsilon x(t)]. \tag{2.19}$$

2.3.2 *Discrete-Time Sliding Mode Control*

In order to take advantage of digital processors, a discrete-time sliding mode control (DSMC) is being used. The DSMC has several advantages over package loss, time delay, handling complex algorithms, and many more [20–24]. Let the system defined in Eq. (2.3) discretized at the sampling time τ can be given as

$$X(k+1) = E(k)X(k) + F(k)u(k)$$

$$Y(k) = G(k)X(k). \tag{2.20}$$

Now, the sliding surface for the discrete-time domain can be given as

$$\mathbb{S}(k) = \Upsilon_d X(k) \tag{2.21}$$

where Υ_d is sliding mode gain which can be calculated using Pole Placement method or LQR approach.

$$\mathbb{S}(k+1) = \Upsilon_d X(k+1) \tag{2.22}$$

$$\mathbb{S}(k+1) = \Upsilon_d(E(k)X(k) + F(k)u(k)) \tag{2.23}$$

Below is a summary of the reaching laws that various researchers from across the world suggested in the discrete-time domain.

- Sarpturk's reaching law [25]

$$|\mathbb{S}(k+1)| < |\mathbb{S}(k)|. \tag{2.24}$$

- Gao's reaching law [26]

$$\mathbb{S}(k+1) = (1-\kappa_1\tau)\mathbb{S}(k) - \kappa_2\tau sgn(\mathbb{S}). \tag{2.25}$$

where τ is sampling rate, $\kappa_1, \kappa_2 > 0$, and $(1-\kappa_1\tau) > 0$.

- Bartoszewicz's reaching law [27]

$$\mathbb{S}(k+1) = d(k) - d(0) + \mathbb{S}_d(k+1), \tag{2.26}$$

where d(k) can be defined as $d_l < d_k < d_h$, d_l is lower bound and d_h is upper bound. $\mathbb{S}_d$ is a priori known function such that the following conditions apply,

If $\mathbb{S}(0) > 2\delta_d$ then

$$\mathbb{S}_d(0) = \mathbb{S}(0)$$

$$\mathbb{S}_d(k)\mathbb{S}_d(0) \geq 0, k \geq 0$$

$$\mathbb{S}_d(k) = 0, k \geq k^*$$

$$|\mathbb{S}_d(k+1)| < |\mathbb{S}_d(k)| - 2\delta_d. \tag{2.27}$$

- Power rate reaching law [28]

$$\mathbb{S}(k+1) = (1-\kappa_1\tau)\mathbb{S}(k) - \kappa_2\tau|\mathbb{S}|^\eta sgn(\mathbb{S}),\ 0 < \eta < 1. \tag{2.28}$$

Now, considering Gao's reaching law in Eq. (2.15), u(k) can be obtained using Eqs. (2.13) and (2.15) as

$$u(k) = -(\Upsilon_d F)^{-1}[E\Upsilon_d X(k) - (1-\kappa_1\tau)\mathbb{S}(k) + \kappa_2\tau sgn(\mathbb{S})]. \tag{2.29}$$

- **Discrete-time Super-twisting Algorithm**

When a closed-loop system is driven into continuous-time sliding mode, a discontinuous control signal switches with an infinite frequency. While in the case of discrete-time domain, the states travel in a zigzag pattern along the surface, which is known as the quasi-sliding mode band (QSMB), but they can never be on the sliding surface. For that DSTA is introduced to reduce the QSMB, more robust and faster the convergence [29].

The switching law of DSTA $u_{sw}(k)$ is given below as [30]

$$u_{sw}(k) = -\kappa_1\sqrt{|\mathbb{S}(k)|}sgn(\mathbb{S}(k)) + \omega(k)$$

$$\omega(k+1) = \omega(k) - \kappa_2\tau sgn(\mathbb{S}(k)). \tag{2.30}$$

where $\kappa_1, \kappa_2 > 0$.

Similarly, the control law u(k) DSTA is a combination of two terms,

$$u(k) = u_{sw}(k) + u_{eq}(k) \tag{2.31}$$

where, $u_{eq}(k)$ is a equivalent control law in discrete-time domain. For finding $u_{eq}(k)$ let's consider $\mathbb{S}(k+1) = 0$ in Eq. (2.23) and derive as

$$u_{eq}(k) = -(\Upsilon_d F)^{-1}[E\Upsilon_d x(k)]. \tag{2.32}$$

2.4 Conclusion

This chapter concludes the overview of the MAS concept, graph theory, and sliding mode control in both continuous-time and discrete-time domains. Nowadays, the MAS concept is being widely used by researchers for the application of consensus, formation, schooling, etc. The graph theory is used for representation of the communication in the MAS.

References

1. Weiss, G. (Ed.): Multiagent systems: a modern approach to distributed artificial intelligence. MIT Press (1999)
2. Multiagent systems. In: Multiagent Systems. Lecture Notes in Computer Science, vol. 799. Springer, Berlin, Heidelberg. https://doi.org/10.1007/BFb0030532
3. Wood, M.F., DeLoach, S.A.: An overview of the multiagent systems engineering methodology. In: Ciancarini, P., Wooldridge, M.J. (eds.) Agent-Oriented Software Engineering. AOSE 2000. Lecture Notes in Computer Science, vol. 1957. Springer, Berlin, Heidelberg (2001). https://doi.org/10.1007/3-540-44564-1_14
4. Jiao, Q., Zhang, H., Xu, S., Lewis, F.L., Xie, L.: Bipartite tracking of homogeneous and heterogeneous linear multi-agent systems. Int. J. Control **92**(12), 2963–2972 (2019)
5. Patel, K., Mehta, A.: Discrete-time sliding mode control for leader following discrete-time multi-agent system. In: IECON 2018-44th Annual Conference of the IEEE Industrial Electronics Society, pp. 2288–2292 (2018)

6. Patel, K.,Mehta, A.: Discrete higher order sliding mode protocol for leader-following consensus of heterogeneous discrete multi-agent system. In: Lecture Notes in Electrical Engineering, pp. 1–10. Springer, Singapore (2019)
7. Patel, K., Mehta, A.: Discrete-time event-triggered higher order sliding mode control for consensus of 2-dof robotic arms. Eur. J. Control (2020)
8. Utkin, V.: Variable structure systems with sliding modes. IEEE Trans. Autom. Control **22**, 212–222 (1977)
9. Utkin, V.I.: Sliding mode control design principles and applications to electric drives. IEEE Trans. Ind. Electron. **40**, 23–36 (1993)
10. Betin, F., Capolino, G.: Sliding mode control for electrical drives. In: 2015 IEEE International Electric Machines Drives Conference (IEMDC), pp. 1043–1048 (2015)
11. Furuta, K., Pan, Y.: Variable structure control with sliding sector. Automatica **36**, 211–228 (2000)
12. Emel'yanov, S.V., Utkin, V.I.: Synthesis of variable structure control systems with a discontinuous switching function. AN SSSR. Tech. Cyber **1**, 182–186 (1964)
13. Young, K.D., Utkin, V.I., Ozguner, U.: A control engineer's guide to sliding mode control. IEEE Trans. Control Syst. Technol. **7**(3), 328–342 (1999). https://doi.org/10.1109/87.761053
14. Shtessel, Y., Edwards, C., Fridman, L., Levant, A.: Sliding Mode Control and Observation, vol. 10. Springer, New York (2014)
15. Mehta, A., Bandyopadhyay, B.: Frequency-Shaped and Observer-Based Discrete-Time Sliding Mode Control. Springer, India (2015)
16. Devika, K.B., Thomas, S.: Power rate exponential reaching law for enhanced performance of sliding mode control. Int. J. Control Autom. Syst. **15**, 2636–2645 (2017)
17. Park, M.-H., Kim, K.-S.: Chattering reduction in the position control of induction motor using the sliding mode. IEEE Trans. Power Electron. **6**, 317–325 (1991)
18. Kawamura, A., Itoh, H., Sakamoto, K.: Chattering reduction of disturbance observer based sliding mode control. IEEE Trans. Ind. Appl. **30**, 456–461 (1994)
19. Kamal, S., Chalanga, A., Moreno, J.A., Fridman, L., Bandyopadhyay, B.: Higher order super-twisting algorithm. In: 13th International Workshop on Variable Structure Systems (VSS), vol. 2014, pp. 1–5. Nantes, France (2014). https://doi.org/10.1109/VSS.2014.6881129
20. Milosavljevic, C.: General condition for the existence of a quasi-sliding mode in the switching hyperplane in discrete variable structure system. Autom. Remote Control **46**, 307–314 (1985)
21. Corless, M., Manela, J.: Control of uncertain discrete-time systems. In: 1986 American Control Conference, pp. 515–520 (1986)
22. Furuta, K.: Sliding mode control of a discrete system. Syst. Control Lett. **14**, 145–152 (1990)
23. Chan, C.: Robust discrete-time sliding mode controller. Syst. Control Lett. **23**(5), 371–374 (1994)
24. Bartolini, G., Ferrara, A., Utkin, V.: Adaptive sliding mode control in discrete-time systems. Automatica **31**(5), 769–773 (1995)
25. Sarpturk, S., Istefanopulos, Y., Kaynak, O.: On the stability of discrete-time sliding mode control systems. IEEE Trans. Autom. Control **32**, 930–932 (1987)
26. Gao, W., Wang, Y., Homaifa, A.: Discrete-time variable structure control systems. IEEE Trans. Ind. Electron. **42**, 117–122 (1995)
27. Bartoszewicz, A.: Discrete-time quasi-sliding-mode control strategies. IEEE Trans. Ind. Electron. **45**, 633–637 (1998)
28. Devika, K.B., Thomas, S.: Power rate exponential reaching law for enhanced performance of sliding mode control. Int. J. Control Autom. Syst. **15**, 2636–2645 (2017)
29. Salgado, I., Kamal, S., Chairez, I., Bandyopadhyay, B., Fridman, L.: Super-twisting-like algorithm in discrete time nonlinear systems. In: The 2011 International Conference on Advanced Mechatronic Systems, Zhengzhou, China, 2011, pp. 497–502 (2011)
30. Modi, A., Joshi, N., Mehta, A.: Robust discrete-time super twisting formation protocol for a 6-DOF Quadcopter swarm. ISA Trans. (2023). ISSN 0019-0578. https://doi.org/10.1016/j.isatra.2023.09.029

Chapter 3
Continuous-Time Sliding Mode Protocol for Formation of Quadcopter Swarm

3.1 System Description of Quadcopter

A Quadcopter is a type of drone that is powered by four rotors, which are arranged in a symmetric pattern around the central body. It has an onboard flight controller that can stabilize at the desired setpoint. The Quadcopter's flight controller uses a combination of sensors, such as accelerometers, gyroscopes, and magnetometers, to measure the orientation, speed, and altitude. These sensors work together to provide real-time feedback to the flight controller, which adjusts the speed of each rotor to maintain stability and control the Quadcopter's movement. The 6-Degrees Of Freedom (DOF) Quadcopter system is shown in Fig. 3.1.

The Quadcopter system consists of the following components:

- Frame: The frame of the Quadcopter is made of lightweight and durable materials, such as carbon fiber or plastic. It has four arms, which are arranged in a cross pattern, and a central platform for mounting the electronic components.
- Motors: The Quadcopter is powered by four brushless DC motors, one for each arm. These motors are high-performance and provide the drone with the necessary lift and maneuverability.
- Propellers: The Quadcopter uses four propellers, one for each motor. These propellers are designed to provide maximum lift and stability to the drone.
- Flight Controller: The flight controller is the brain of the Quadcopter. It is a small electronic board that receives data from various sensors, such as accelerometers and gyroscopes uses this data to control the motors and stabilize the drone in flight.
- Battery: The Quadcopter is powered by a high-capacity lithium-polymer battery. The battery is mounted on the central platform of the drone and provides power to the motors and electronics.
- Radio Transmitter and Receiver: The Quadcopter is controlled by a radio transmitter and receiver. The transmitter is held by the user and sends signals to the receiver on the Quadcopter, which in turn controls the motors and flight controller.

A. Mehta and A. Modi, *Robust Sliding Mode Protocols for Formation of Quadcopter Swarm*, Studies in Systems, Decision and Control 521,
https://doi.org/10.1007/978-981-99-9726-8_3

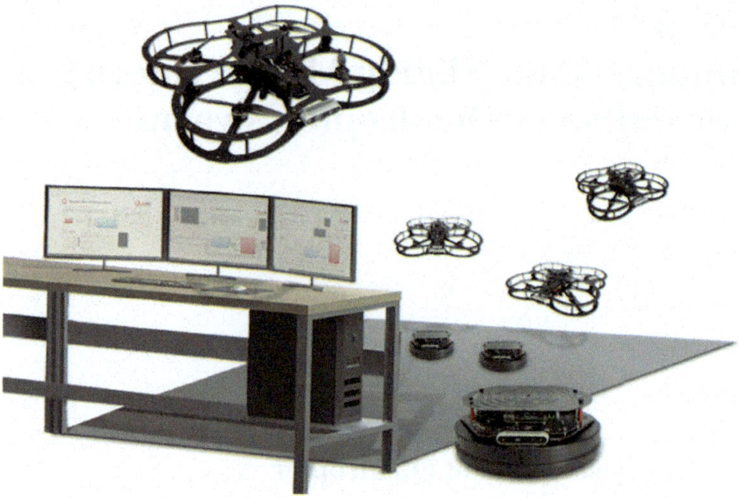

Fig. 3.1 Quadcopter swarm. Courtesy: Quanser PVT. LTD

3.2 Distributed Continuous-Time Sliding Mode (DCSM) Formation Protocol for Linear Model of Quadcopter Swarm

Figure 3.2 shows the block diagram for the formation of the Quadcopter swarm. As per the block diagram, the reference is provided to the primary control of the leader agent which will get the Quadcopter to the desired location. All agents shared their

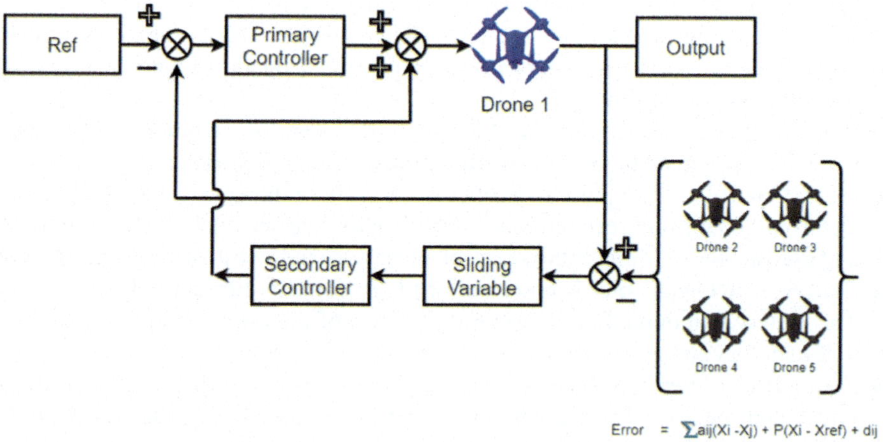

Fig. 3.2 Block diagram for formation of Quadcopter swarm

information to the neighborhood agent as per the formation pattern shown in Fig. 3.7. The summation of these information is known as the local error which is given in Eq. (3.28). Only the leader Quadcopter has access to the reference trajectory. Then the secondary controller is maintaining the desired distance between all the agents which will further add to the system through the primary controller.

In order to achieve the formation, let us derive the mathematical model using the kinematics and dynamics of the system. The mathematical model of the Quadcopter is derived in Sect. 3.2.1.

3.2.1 Linear Model of Quadcopter

A linear model of a Quadcopter represents the system's behavior using linear equations, which is an approximation of the real-time dynamics. According to the working principle, ω_1 and ω_3 rotate in a clockwise direction, while ω_2 and ω_4 rotate in an anticlockwise direction [1]. The difference between ω_1 and ω_3 provides the torque τ_x about the X-axis.

$$\tau_x = (\omega_1^2 - \omega_3^2)K_t l \tag{3.1}$$

The difference between ω_2 and ω_3 provide the torque τ_y about the Y-axis.

$$\tau_y = (\omega_2^2 - \omega_4^2)K_t l \tag{3.2}$$

The difference between clockwise and anticlockwise torque provides the τ_z about the Z-axis.

$$\tau_z = (\omega_2^2 + \omega_4^2 - \omega_1^2 - \omega_3^2)K_d \tag{3.3}$$

As per the above-given equations, the dynamics of the Quadcopters are calculated below:

$$m\ddot{X} = (F_1 + F_2 + F_3 + F_4)(c\phi s\theta c\psi + s\phi s\psi) \tag{3.4}$$

$$m\ddot{Y} = (F_1 + F_2 + F_3 + F_4)(c\phi s\theta c\psi + s\phi c\psi) \tag{3.5}$$

$$m\ddot{Z} = (F_1 + F_2 + F_3 + F_4)(c\phi c\theta) - mg \tag{3.6}$$

$$I_x\ddot{\phi} = (F_1 - F_3)l + \dot{\theta}\dot{\psi}(I_y - I_z) \tag{3.7}$$

$$I_y\ddot{\theta} = (F_2 - F_4)l + \dot{\psi}\dot{\phi}(I_z - I_x) \tag{3.8}$$

$$I_z\ddot{\psi} = (M_2 + M_4 - M_1 - M_3) + \dot{\theta}\dot{\phi}(I_x - I_y) \tag{3.9}$$

Table 3.1 System parameters of Quadcopter

Parameter	Value
Weight	1.2 kg
Maximum payload	1.2 kg
Flight time	15–20 s
Maximum speed	54 m/s
Maximum altitude	4000 m
Maximum range	1 km
Battery(LiPo)	2200 mAh
Motors(Brushless DC Motors)	920 KV
Propellers	10 × 4.5 inch
Flight controller	Pixhawk
I_x Roll moment of inertia	0.11 kgm^2
I_y Pitch moment of inertia	0.11 kgm^2
I_z Yaw moment of inertia	0.04 kgm^2
K_t thrust co-efficient	3×10^{-6}
K_d Drag co-efficient	4×10^{-9}

where F = $(F_1 + F_2 + F_3 + F_4)$. m is mass. M_i is Momentum for respective motor, i = 1, 2, 3, 4 and I is the Inertia for respective direction. m is the mass of the Quadcopter.

The dynamics mentioned above are non-linear in nature. So for controlling the system, their dynamics must be in a linearized form which is given below

$$\ddot{X} = \frac{F}{m}\theta; \quad \ddot{\phi} = \frac{\tau_x}{I_x} \tag{3.10}$$

$$\ddot{Y} = \frac{F}{m}\phi; \quad \ddot{\theta} = \frac{\tau_y}{I_y} \tag{3.11}$$

$$\ddot{Z} = \frac{F}{m} - g; \quad \ddot{\psi} = \frac{\tau_z}{I_z} \tag{3.12}$$

The values of Quadcopter constants are mentioned in Table 3.1.

Now to design the primary control of Quadcopter, the linear quadratic regulator (LQR) method is proposed using state space approach. The ideal state space model of Quadcopter in continuous-time domain is given in Eq. (3.13).

$$\dot{X}(t) = Ax(t) + Bu(t)$$

$$Y(t) = Cx(t) \tag{3.13}$$

where A(t) and B(t) are the system and input matrices. Also x(t), u(t), and Y(t) are the state, input, and output vectors, respectively. Hence, the state space model for the system is given below

$$\dot{X} = \begin{bmatrix} 0&1&0&0&0&0&0&0&0&0&0&0 \\ 0&0&0&0&0&0&0&g&0&0&0&0 \\ 0&0&0&1&0&0&0&0&0&0&0&0 \\ 0&0&0&0&0&0&g&0&0&0&0&0 \\ 0&0&0&0&0&1&0&0&0&0&0&0 \\ 0&0&0&0&0&0&0&0&0&0&0&0 \\ 0&0&0&0&0&0&0&1&0&0&0&0 \\ 0&0&0&0&0&0&0&0&0&0&0&0 \\ 0&0&0&0&0&0&0&0&0&1&0&0 \\ 0&0&0&0&0&0&0&0&0&0&0&0 \\ 0&0&0&0&0&0&0&0&0&0&0&1 \\ 0&0&0&0&0&0&0&0&0&0&0&0 \end{bmatrix} \begin{bmatrix} x \\ \dot{x} \\ y \\ \dot{y} \\ z \\ \dot{z} \\ \theta \\ \dot{\theta} \\ \phi \\ \dot{\phi} \\ \psi \\ \dot{\psi} \end{bmatrix} + \begin{bmatrix} 0&0&0&0 \\ 0&0&0&0 \\ 0&0&0&0 \\ 0&0&0&0 \\ 0&0&0&0 \\ \frac{1}{m}&0&0&0 \\ 0&0&0&0 \\ 0&\frac{1}{I_x}&0&0 \\ 0&0&0&0 \\ 0&0&\frac{1}{I_y}&0 \\ 0&0&0&0 \\ 0&0&0&\frac{1}{I_z} \end{bmatrix} \begin{bmatrix} F \\ \tau_x \\ \tau_y \\ \tau_z \end{bmatrix} \quad (3.14)$$

$$Y = \begin{bmatrix} x \\ y \\ z \\ \theta \\ \phi \\ \psi \end{bmatrix} = \begin{bmatrix} 1&0&0&0&0&0&0&0&0&0&0&0 \\ 0&0&1&0&0&0&0&0&g&0&0&0 \\ 0&0&0&0&1&0&0&0&0&0&0&0 \\ 0&0&0&0&0&0&1&0&0&0&0&0 \\ 0&0&0&0&0&0&0&0&1&0&0&0 \\ 0&0&0&0&0&0&0&0&0&0&1&0 \end{bmatrix} \begin{bmatrix} x \\ \dot{x} \\ y \\ \dot{y} \\ z \\ \dot{z} \\ \theta \\ \dot{\theta} \\ \phi \\ \dot{\phi} \\ \psi \\ \dot{\psi} \end{bmatrix} + \begin{bmatrix} 0&0&0&0 \\ 0&0&0&0 \\ 0&0&0&0 \\ 0&0&0&0 \end{bmatrix} \begin{bmatrix} F \\ \tau_x \\ \tau_y \\ \tau_z \end{bmatrix} \quad (3.15)$$

where A(t) $\in$ 12×12, B(t) $\in$ 12×4, C(t) $\in$ 12×4, x(t) $\in$ 12×1, u(t) $\in$ 4×1.

$$x(t) = \begin{bmatrix} x_1 \\ x_2 \\ x_3 \\ x_4 \\ x_5 \\ x_6 \\ x_7 \\ x_8 \\ x_9 \\ x_{10} \\ x_{11} \\ x_{12} \end{bmatrix} = \begin{bmatrix} x \\ \dot{x} \\ y \\ \dot{y} \\ z \\ \dot{z} \\ \theta \\ \dot{\theta} \\ \phi \\ \dot{\phi} \\ \psi \\ \dot{\psi} \end{bmatrix} = \begin{bmatrix} Distance\ in\ X-axis \\ Velocity\ in\ X-axis \\ Distance\ in\ Y-axis \\ Velocity\ in\ Y-axis \\ Distance\ in\ Z-axis \\ Velocity\ in\ Z-axis \\ Angular\ displacement\ around\ X-axis \\ Rate\ of\ change\ of\ Angular\ displacement\ in\ X-axis \\ Angular\ displacement\ around\ Y-axis \\ Rate\ of\ change\ of\ Angular\ displacement\ in\ Y-axis \\ Angular\ displacement\ around\ Z-axis \\ Rate\ of\ change\ of\ Angular\ displacement\ in\ Z-axis \end{bmatrix} \quad (3.16)$$

3.2.2 Continuous-Time Primary Control for Linear Model of Quadcopter

Primary controller is a device or software that regulates the process variable to maintain a desired setpoint for the Quadcopter. The primary controller receives input from the feedback sensors that measure the parameters and adjusts its output accordingly to maintain the position. Here, linear quadratic regulator (LQR) technique is used to design the primary controller. In the LQR approach, we need to design the feedback gain adequately, which will result in the Quadcopter becoming stable at the desired position.

This feedback gain can be calculated using the cost function of the LQR control given below [2]:

$$J = \frac{1}{2}\int_0^{\infty} (x^T Qx + u^T Ru)dx \tag{3.17}$$

where Q and R are the weighted matrices depending on the required control effort to stabilize the Quadcopter. The parameters of the Q matrix will depend on the states of the Quadcopter and R will depend on the input of the Quadcopter.

$$Q = \begin{bmatrix} 1\,0\,0\,0\,0\,0\,0\,0\,0\,0\,0\,0 \\ 0\,0\,0\,0\,0\,0\,0\,0\,0\,0\,0\,0 \\ 0\,0\,1\,0\,0\,0\,0\,0\,0\,0\,0\,0 \\ 0\,0\,0\,0\,0\,0\,0\,0\,0\,0\,0\,0 \\ 0\,0\,0\,0\,1\,0\,0\,0\,0\,0\,0\,0 \\ 0\,0\,0\,0\,0\,0\,0\,0\,0\,0\,0\,0 \\ 0\,0\,0\,0\,0\,0\,1\,0\,0\,0\,0\,0 \\ 0\,0\,0\,0\,0\,0\,0\,0\,0\,0\,0\,0 \\ 0\,0\,0\,0\,0\,0\,0\,0\,1\,0\,0\,0 \\ 0\,0\,0\,0\,0\,0\,0\,0\,0\,0\,0\,0 \\ 0\,0\,0\,0\,0\,0\,0\,0\,0\,0\,1\,0 \\ 0\,0\,0\,0\,0\,0\,0\,0\,0\,0\,0\,0 \end{bmatrix}; \quad R = \begin{bmatrix} 1\,0\,0\,0 \\ 0\,1\,0\,0 \\ 0\,0\,1\,0 \\ 0\,0\,0\,1 \end{bmatrix} \tag{3.18}$$

As per the system matrix A(t) and the input matrix u(t), the dimension of the Q matrix will be 12×12 and the dimension of the R matrix is 4×4. The control law for the primary controller is given in Eq. (3.19),

$$u = -K_{Cont}x(t) \tag{3.19}$$

where K_{Cont} is the feedback gain having the dimension of 4×12, which can be calculated using Q and R weighted matrices.

3.2.3 Simulation Results of Continuous-Time Primary Control for Linear Model of Quadcopter

In this section, the simulation results are discussed for the primary control for the linearized model of the Quadcopter. The simulation results are carried out using the MATLAB/Simulink software. As per the block diagram given in Fig. 3.2, the reference vector given to the Quadcopter has the dimension of 12×1. Hence, the reference is given to the Quadcopter as $[10\ 0\ 5\ 0\ 10\ 0\ 0\ 0\ 0\ 0\ 0\ 0]^{\tau}$. The feedback gain K_{Cont} of 4×12 matrix is obtained from the LQR approach given as

$$K_{Cont} = \begin{bmatrix} 0 & 0 & 0 & 0 & 1 & 0 & 0 & 0 & 0 & 0 & 0 & 0 \\ 0 & 0 & 1 & 0.88 & 0 & 0 & 3.85 & 0.92 & 0 & 0 & 0 & 0 \\ 1 & 0.88 & 0 & 0 & 0 & 0 & 0 & 0 & 3.85 & 0.92 & 0 & 0 \\ 0 & 0 & 0 & 0 & 0 & 0 & 0 & 0 & 0 & 0 & 0.99 & 0.28 \end{bmatrix} \tag{3.20}$$

Figure 3.3 shows the transitional (X, Y, & Z) motion and their velocity (rate of change of position) of the Quadcopter. The Quadcopter becomes stable in the desired

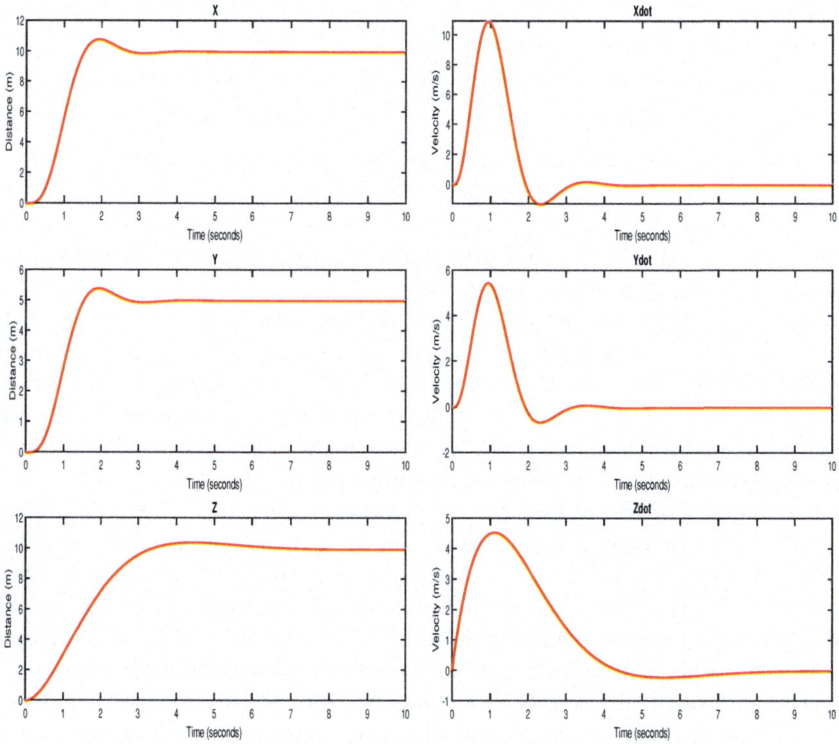

Fig. 3.3 Transitional motion of continuous-time primary control of Quadcopter

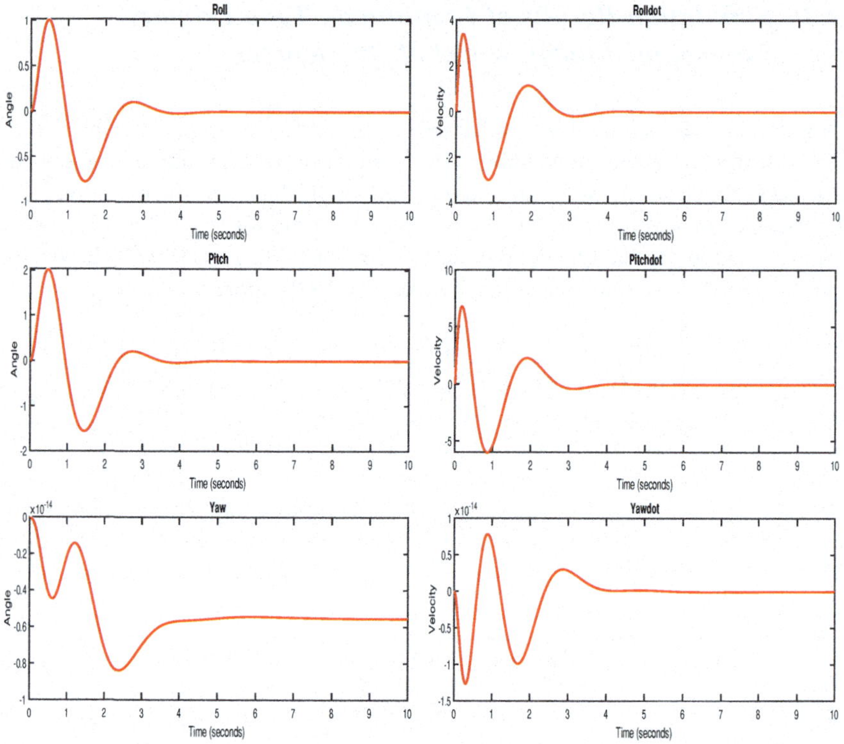

Fig. 3.4 Rotational motion of continuous-time primary control of Quadcopter

location, i.e., $[X\ Y\ Z]^\tau = [10\ 5\ 10]^\tau$. Once the Quadcopter reaches the desired position, the velocity also becomes zero.

Figure 3.4 is the rotational (Roll, Pitch, & Yaw) motion and the angular velocity(rate of change of angular position) of the Quadcopter. Figure 3.5 shows the required control effort which refers to the amount of energy or force applied to the system's input in order to achieve the desired output or maintain the system's stability and performance. Here, one trajectory path is also given which is followed by the Quadcopter as per the given input shown in Fig. 3.6.

Distributed Control of Quadcopter Swarm: A distributed controller referred to as the formation control is responsible for the global consensus between all the Quadcopters for a given pattern. In order to achieve the formation Quadcopter swarm, the graph theory is used to establish the communication between all the Quadcopters, and MAS is used to design the global formation control. The overview of MAS and graph theory is given in Sects. 2.1 and 2.2. MAS is a group of multiple agents having the same or different characteristics and functionality. Mostly, the MAS application is used for those goals which are difficult or impossible for a single agent.

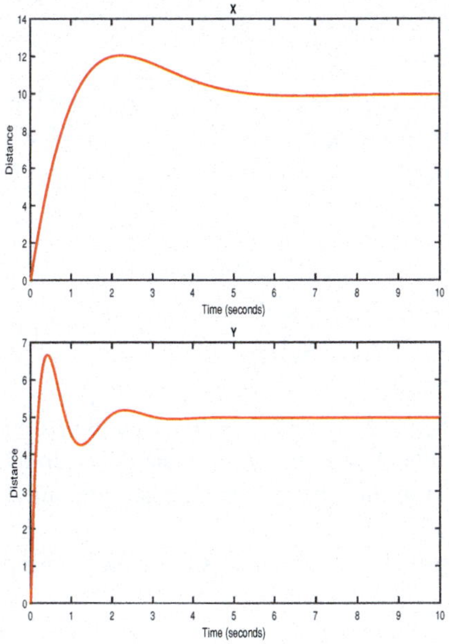

Fig. 3.5 Control efforts of continuous-time primary control of Quadcopter

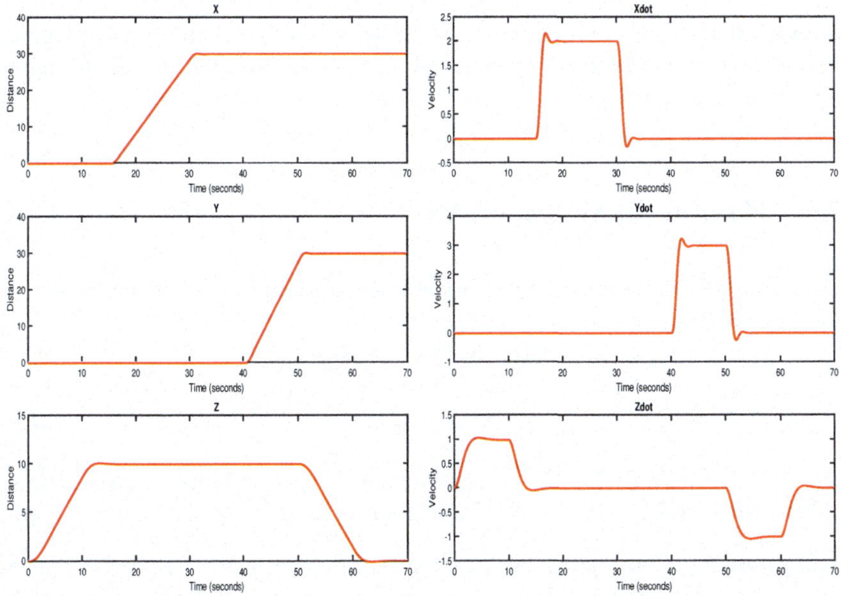

Fig. 3.6 Trajectory tracking of continuous-time primary control of Quadcopter

3.2.4 Graph Theory

Let us consider 5 number of Quadcopters connected and the undirected graph theory is used for communicating in both directions. The graph $\mathbb{G}$ is defined as $\mathbb{G} = (\mathbb{V}, \mathcal{E}, \mathbb{A})$ where $\mathbb{V} = [1, 2, 3, 4, 5]$ shows the number of nodes present in the graph and $\mathcal{E} \subseteq \mathbb{V} \times \mathbb{V}$ indices the edges between the nodes [3].

Adjacency matrix $\mathbb{A}$ is defined as

$$\mathbb{A} = \left[a_{ij}\right] : \begin{cases} a_{ij} = 1 & \text{if}(i, j)\epsilon\mathcal{E} \\ a_{ij} = 0 & \text{otherwise} \end{cases} \tag{3.21}$$

The Laplacian matrix $\mathbb{L}$ is defined as $\mathbb{L} = \mathbb{D} - \mathbb{A}$, where $\mathbb{D}$ is $m \times m$ square degree matrix which shows the number of edges connected to the agent. When the Laplacian matrix $\mathbb{L}$ of a directed graph $\mathbb{G}$ possesses a unique zero eigenvalue accompanied by its corresponding eigenvector $1_5 = [1, 1, 1, 1, 1]^\tau$.

The pinning gain matrix $\mathbb{P}$ is 4×4 diagonal matrix which shows the pinning of the reference with the nodes.

$$\mathbb{P} = diag\left[p_i\right] : \begin{cases} p_i > 0 & \text{if} V_{ref} \text{is pinned} \\ p_i = 0 & \text{otherwise} \end{cases} \tag{3.22}$$

Assumption 3.1 Considering the $(\mathbb{L} + \mathbb{P})$ is a positive definite and symmetric matrix if at least one follower receives information from the leader through the connections in graph $\mathbb{G}$ [4].

3.2.5 Multi-agent System (MAS)

Considering the continuous-time identical linear system for the ith agent as

$$\dot{x}_i(t) = Ax_i(t) + Bu_i(t)$$

$$y_i(t) = Cx_i(t) \tag{3.23}$$

where $A \in \mathbb{R}^{12\times 12}$ and $B \in \mathbb{R}^{12\times 4}$ are the plant matrix and input matrix. The state vector $x_i(t) = [x_1(t)\, x_2(t) \ldots.\, x_{12}(t)]^\tau$ and input vector $u_i(t) = [u_1(t)\, u_2(t) \ldots.\, u_4(t)]^\tau$.

Assumption 3.2 Considering the given system matrix pair(A, B) for ith agent is controllable and the MAS under consideration does have a stationary, non-switching communication topology [5].

From Eq. (3.23), a global MAS is derived as

$$\dot{X}(t) = (I_N \otimes A)X(t) + (I_N \otimes B)u(t) \tag{3.24}$$

we may rewrite Eq. (3.25) as

$$\dot{X}(t) = \bar{A}X(t) + \bar{B}u(t) \tag{3.25}$$

where ⊗ denotes the Kronecker product, N = 5, $\bar{A} = I_5 \otimes A$, and $\bar{B} = I_5 \otimes B$.

Formation Protocol for MAS: For a Quadcopter system, there are four controlling parameters at the user end such as X, Y, Z, and Yaw position. Hence, the formation can be achieved using the linear parameters X, Y, and Z position. Consider a MAS of N Quadcopter agents. Let the desired formation of the agents represented as $\mathbb{F} = (\mathbb{G}, \mathbb{Q})$ where $\mathbb{G}$ is the formation graph and $\mathbb{Q} = [q_1\ q_2\ q_3\ q_4\ q_5]$ [6]. In the multiple Quadcopters system, the formation condition of the Z position between agents i and j can be defined as

$$||Z_i - Z_j|| = d_{ij} \tag{3.26}$$

where d_{ij} is the desired distance between agents i and j. Similarly, the X and Y positions can also be defined.

To design the distributed continuous-time sliding mode (DCSM) formation protocol, the pattern of five Quadcopters as shown in Fig. 3.7 is considered. As per the formation pattern, the position of the leader agent is considered as the (0, 0, 0) for the (X, Y, Z) position. The immediate follower-1 (FA-1) and follower-2 (FA-2) have the distance d_{ij} with the leader agent in (X, Y, Z) position as (–2, 2, 0) and (2, 2, 0). Then the next follower-3 (FA-3) and follower-4 (FA-4) have the distance d_{ij} with leader agent in (X, Y, Z) position as (–2, –2, 0) and (2, –2, 0). The altitude of all the Quadcopters will remain the same or it can be presented as $d_{ij} = 0$ for all Quadcopters in the Z position.

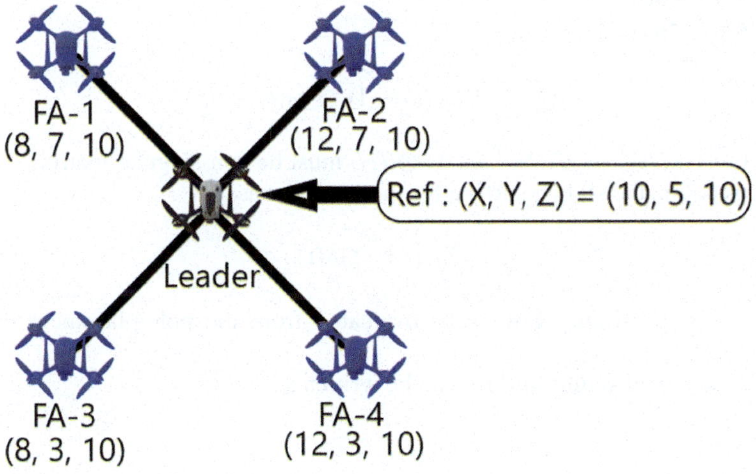

Fig. 3.7 Communication topology

3.2.6 DCSM Formation Protocol for Linear Model of Quadcopter Swarm Using Constant-Proportional Reaching Law

In this section, a distributed control for the leader-follower formation of a linearized model of a Quadcopter swarm is designed using the DCSM protocol using constant-proportional reaching law defined in Theorem 3.1.

Theorem 3.1 *The global homogeneous MAS given in Eq. (3.25) is said to achieve the leader-follower formation with the distributed continuous-time sliding mode (DCSM) protocol using constant-proportional reaching law given as*

$$u(t) = [-(\bar{\Upsilon}\Gamma\bar{B}(t))^{-1}[\bar{\Upsilon}\Gamma\bar{A}(t)X(t) - \bar{\Upsilon}\Gamma(1_5 \otimes \dot{x}_0(t) + \bar{d}_{ij}) + \Lambda_1\bar{\mathbb{S}}(t) + \Lambda_2 sgn(\bar{\mathbb{S}}(t))]] \tag{3.27}$$

Proof The local neighborhood formation error for the MAS is

$$\delta(t) = \sum a_{ij}[[x_i(t) - x_j(t)] + \mathbb{P}_i[x_i(t) - x_0(t)] + d_{ij}] \tag{3.28}$$

where x_i and x_j are the states of the i and j agents, respectively. P_i is the pinning gain and d_{ij} is the distance between the agents. From Eq. (3.28), the formation error in global form can be written as

$$\mathbb{E}(t) = (((I_5 + \bar{\mathbb{D}} + \bar{\mathbb{B}})^{-1})(\bar{\mathbb{L}} + \bar{\mathbb{B}}) \otimes I_{12})(X(t) - 1_5 \otimes x_0(t) + \bar{d}_{ij}) \tag{3.29}$$

where $\bar{\mathbb{D}}$, $\bar{\mathbb{B}}$ and $\bar{\mathbb{L}}$ are degree, diagonal and Laplacian matrices, respectively. Also, I_5 = diag(5), I_{12} = diag(12), $1_5 = [1, 1, 1, 1, 1]^\tau$ and $\bar{d}_{ij}$ is a global form of the distance between all agents.

So, Eq. (3.29) is written as

$$\mathbb{E}(t) = \Gamma(X(t) - 1_5 \otimes x_0(t) + \bar{d}_{ij}) \tag{3.30}$$

where $\Gamma = ((I_5 + \bar{\mathbb{D}} + \bar{\mathbb{B}})^{-1})(\bar{\mathbb{L}} + \bar{\mathbb{B}}) \otimes I_{12})$ must be non-singular matrix.

Let us define the sliding surface for the ith agent as

$$\mathbb{S}_i(t) = \Upsilon_i\delta(t) \tag{3.31}$$

where Υ_i is the sliding gain to be evaluated from the pole-placement or LQR approach.

Now the global sliding surface can be written as

$$\bar{\mathbb{S}}(t) = \bar{\Upsilon}\mathbb{E}(t) \tag{3.32}$$

where $\bar{\Upsilon} = I_{12} \otimes \Upsilon_i$.

Using Eq. (3.32), we may write

$$\bar{\mathbb{S}}(t) = \bar{\Upsilon}\Gamma(X(t) - 1_5 \otimes x_0(t) + \bar{d}_{ij}) \tag{3.33}$$

Further, advancing Eq. (3.33)

$$\dot{\bar{\mathbb{S}}}(t) = \bar{\Upsilon}\Gamma(\dot{X}(t) - 1_5 \otimes \dot{x}_0(t) + \bar{d}_{ij}) \tag{3.34}$$

Substituting Eq. (3.25) into Eq. (3.34), we may get

$$\dot{\bar{\mathbb{S}}}(t) = \bar{\Upsilon}\Gamma(\bar{A}X(t) + \bar{B}u(t) - 1_5 \otimes \dot{x}_0(t) + \bar{d}_{ij}) \tag{3.35}$$

Let us define the constant-proportional reaching law as [7, 8]

$$\dot{\mathbb{S}}(t) = -\Lambda_1\mathbb{S}(t) - \Lambda_2 sgn(\mathbb{S}(t)) \tag{3.36}$$

where Λ_1, $\Lambda_2 > 0$. The constant-proportional reaching law can be written in the global form as

$$\dot{\bar{\mathbb{S}}}(t) = -\bar{\Lambda}_1\bar{\mathbb{S}}(t) - \bar{\Lambda}_2 sgn(\bar{\mathbb{S}}(t)) \tag{3.37}$$

Now, comparing Eqs. (3.35) and (3.37), we may get the DCSM formation protocol for the linear model of Quadcopter swarm using constant-proportional reaching law as

$$u(t) = [-(\bar{\Upsilon}\Gamma\bar{B}(t))^{-1}[\bar{\Upsilon}\Gamma\bar{A}(t)X(t) - \bar{\Upsilon}\Gamma(1_5 \otimes \dot{x}_0(t) + \bar{d}_{ij}) + \bar{\Lambda}_1\bar{\mathbb{S}}(t) + \bar{\Lambda}_2 sgn(\bar{\mathbb{S}}(t))]] \tag{3.38}$$

This completes the proof.

The next presents the global stability of a Quadcopter swarm using DCSM protocol using constant-proportional reaching law.

Theorem 3.2 *The global stability of the leader-following formation of Quadcopter swarm using DCSM constant-proportional reaching law given in Eq. (3.38) is said to be stable, if the closed-loop error dynamics in Eq. (3.29) of MAS drives toward the global sliding surface and maintained on it for any gain* Λ_1 *&* $\Lambda_2 > 0$*, provided the following condition is true:*

$$0 \leq \bar{\beta} \tag{3.39}$$

where

$$\bar{\beta} = \bar{\Lambda}_2\bar{\mathbb{S}}(t) sign(\bar{\mathbb{S}}(t)) \tag{3.40}$$

Proof The global system of Quadcopters is given below

$$\dot{X}(t) = \bar{A}X(t) + \bar{B}u(t) \tag{3.41}$$

Firstly, define the global sliding surface as shown in Eq. (3.33)

$$\bar{\mathbb{S}}(t) = \bar{\Upsilon}\Gamma(X(t) - 1_5 \otimes x_0(t) + \bar{d}_{ij}) \tag{3.42}$$

Further, advancing Eq. (3.42)

$$\dot{\bar{\mathbb{S}}}(t) = \bar{\Upsilon}\Gamma(\dot{X}(t) - 1_5 \otimes \dot{x}_0(t) + \bar{d}_{ij}) \tag{3.43}$$

Let us define the Lyapunov function as

$$\mathcal{V}(t) = \frac{1}{2}\bar{\mathbb{S}}^2(t) \tag{3.44}$$

differentiating Eq. (3.44)

$$\dot{\mathcal{V}}(t) = \bar{\mathbb{S}}(t)\dot{\bar{\mathbb{S}}}(t) \tag{3.45}$$

Substituting $\dot{\bar{\mathbb{S}}}(t)$ in Eq. (3.45), we may get

$$\dot{\mathcal{V}}(t) = \bar{\mathbb{S}}(t)\bar{\Upsilon}\Gamma(\bar{A}X(t) + \bar{B}u(t) - 1_5 \otimes \dot{x}_0(t) + \bar{d}_{ij}) \tag{3.46}$$

Substituting the u(t) in Eq. (3.38) into Eq. (3.46), we may get

$$\dot{\mathcal{V}}(t) = \bar{\mathbb{S}}(t)(-\bar{\Lambda}_1\bar{\mathbb{S}}(t) - \bar{\Lambda}_2 sgn(\bar{\mathbb{S}}(t))) \tag{3.47}$$

where $\bar{\Lambda}_1$ & $\bar{\Lambda}_2 > 0$. we may write Eq. (3.47) as

$$\dot{\mathcal{V}}(t) = -\bar{\Lambda}_1\bar{\mathbb{S}}^2(t) - \bar{\Lambda}_2\bar{\mathbb{S}}(t)sgn(\bar{\mathbb{S}}(t)) \tag{3.48}$$

Now denoting,

$$\bar{\beta} = \bar{\Lambda}_2\bar{\mathbb{S}}(t)sign(\bar{\mathbb{S}}(t)) \tag{3.49}$$

Now, we may write the Lyapunov function for the DCSM formation protocol using constant-proportional reaching law as

$$\dot{\mathcal{V}}(t) = -\bar{\beta} - \bar{\Lambda}_1\bar{\mathbb{S}}^2(t) \tag{3.50}$$

For stability purpose, the following condition holds true:

$$0 \leq \bar{\beta} \tag{3.51}$$

From Eq. (3.51), the global stability condition of the DST formation protocol can be obtained by tuning the $\bar{\beta}$ so that the $\dot{\mathcal{V}}(t) < 0$.

This completes the proof.

3.2.7 Simulation Results of DCSM Formation Protocol Using Constant-Proportional Reaching Law

In this section, simulation results of the formation of five Quadcopters are discussed using the DCSM protocol using constant-proportional reaching law. The block diagram of the Quadcopter swarm is given in Fig. 3.2. All agents shared their information to the neighborhood agent as per the formation pattern shown in Fig. 3.7. The summation of these information is known as the local error which is given in Eq. (3.28). Only the leader Quadcopter has access to the reference trajectory. Further this local error can be written in a global form as mentioned in Eq. (3.29). The required parameter to evaluate the global error such as the Laplacian $\mathbb{L}$, adjacency $\mathbb{A}$, degree $\mathbb{D}$, and pinning matrices $\mathbb{P}$ are derived as

$$\mathbb{L} = \begin{bmatrix} 2 & 0 & 0 & 0 & 0 \\ -1 & 2 & 0 & 0 & 0 \\ -1 & 0 & 2 & 0 & 0 \\ 0 & -1 & 0 & -1 & 0 \\ 0 & 0 & -1 & 0 & 1 \end{bmatrix} \tag{3.52}$$

$$\mathbb{D} = \begin{bmatrix} 2 & 0 & 0 & 0 & 0 \\ 0 & 2 & 0 & 0 & 0 \\ 0 & 0 & 2 & 0 & 0 \\ 0 & 0 & 0 & 1 & 0 \\ 0 & 0 & 0 & 0 & 1 \end{bmatrix} \tag{3.53}$$

$$\mathbb{P} = \begin{bmatrix} 1 & 0 & 0 & 0 & 0 \\ 0 & 0 & 0 & 0 & 0 \\ 0 & 0 & 0 & 0 & 0 \\ 0 & 0 & 0 & 0 & 0 \\ 0 & 0 & 0 & 0 & 0 \end{bmatrix} \tag{3.54}$$

$$\mathbb{A} = \begin{bmatrix} 0 & 0 & 0 & 0 & 0 \\ 1 & 0 & 0 & 0 & 0 \\ 1 & 0 & 0 & 0 & 0 \\ 0 & 1 & 0 & 0 & 0 \\ 0 & 0 & 1 & 0 & 1 \end{bmatrix} \tag{3.55}$$

The sliding mode gain Υ_i for the ith Quadcopter agent is designed using the LQR approach. The super-twisting constant Λ_1 and Λ_2 are chosen as 5 and 30, respectively. The output in the direction of X-, Y-, and Z-axis for all agents are shown in Figs. 3.8, 3.9, and 3.10, respectively. The X-axis position of the leader is 10m from the origin. As the d_{ij} is 2m for FA-2 and FA-4 from the leader, the X position of FA-2 and FA-4 is 12m. Similarly d_{ij} is -2m for FA-1 and FA-3, so the X position of FA-1 and FA-3 is 8m. The d_{ij} is 2m for FA-1 and FA-2 from the leader, the Y position of FA-1 and

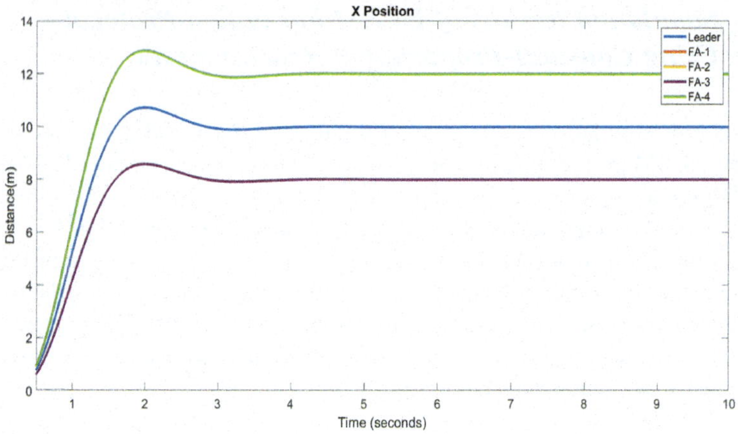

Fig. 3.8 X position of all agents using DCSM constant-proportional reaching law

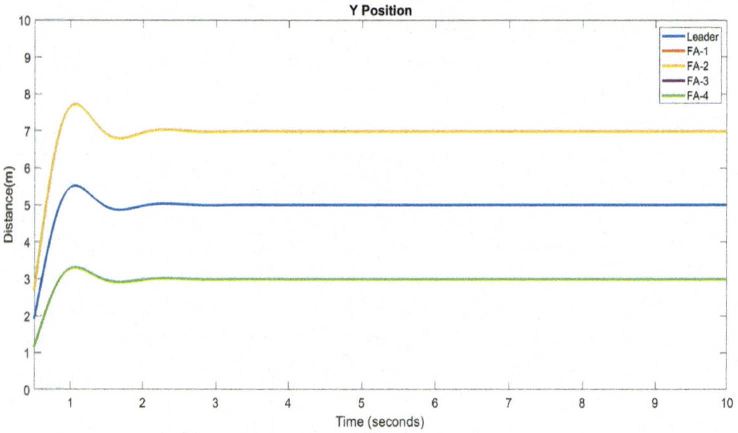

Fig. 3.9 Y position of all agents using DCSM constant-proportional reaching law

FA-2 is 7m. Similarly d_{ij} is -2m for FA-3 and FA-4, the Y position of FA-3 and FA-4 is 3m. And the d_{ij} is zero for the Z position, the altitude of all agents is the same at 10 m.

Figure 3.11 presents the tracking error of X, Y, Z, and Yaw positions for all agents. Figure 3.12 shows the total control effort required for achieving the formation of linear model of Quadcopter swarm using DCSM protocol. The required control band varies between the range of –1 to +1. Figure 3.13 shows the sliding surface of the Quadcopter swarm. The chattering band of the sliding surface is -2×10^{-4} to $+2 \times 10^{-4}$ given in the magnified window.

The sliding mode formation protocol given in Sect. 3.2.6 is derived using the constant-proportional reaching law which is a first-order sliding mode control. The

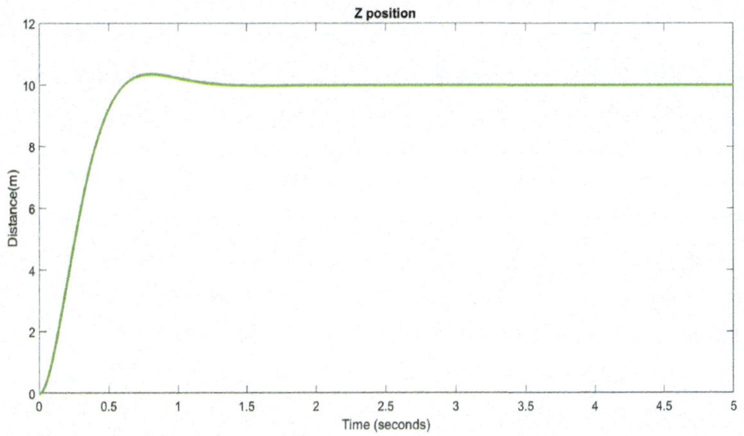

Fig. 3.10 Z position of all agents using DCSM constant-proportional reaching law

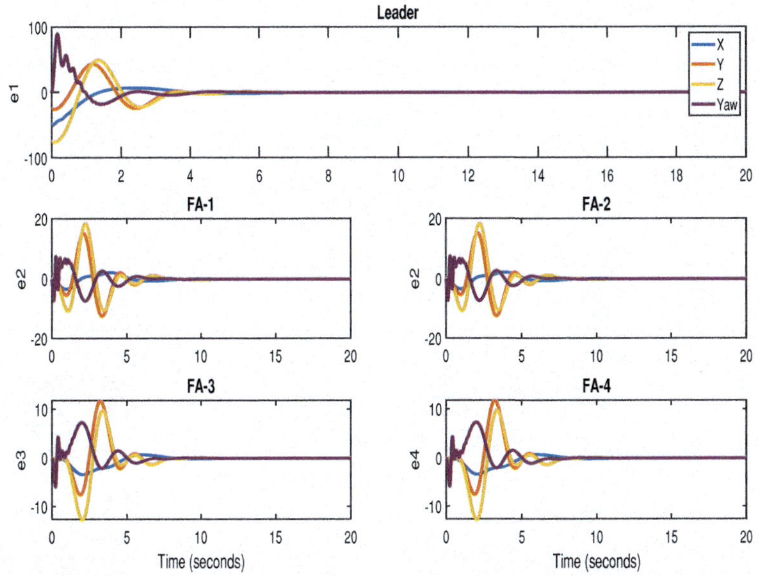

Fig. 3.11 Tracking error of all agents using DCSM constant-proportional reaching law

first-order continuous-time sliding mode protocols have the significant amount of chattering shown in Figs. 3.12 and 3.13. To reduce the chattering effect during the formation, the second-order sliding mode control, namely, super-twisting algorithm is designed.

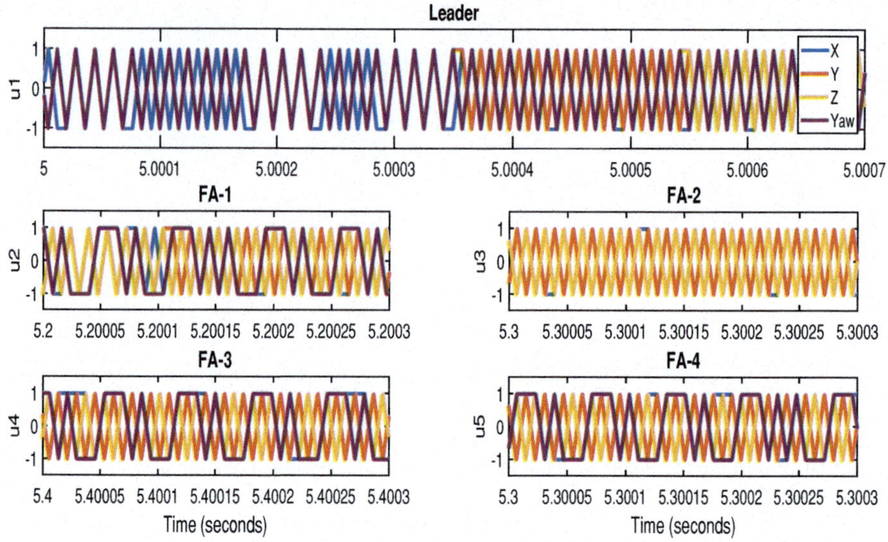

Fig. 3.12 Control efforts of all agents using DCSM constant-proportional reaching law

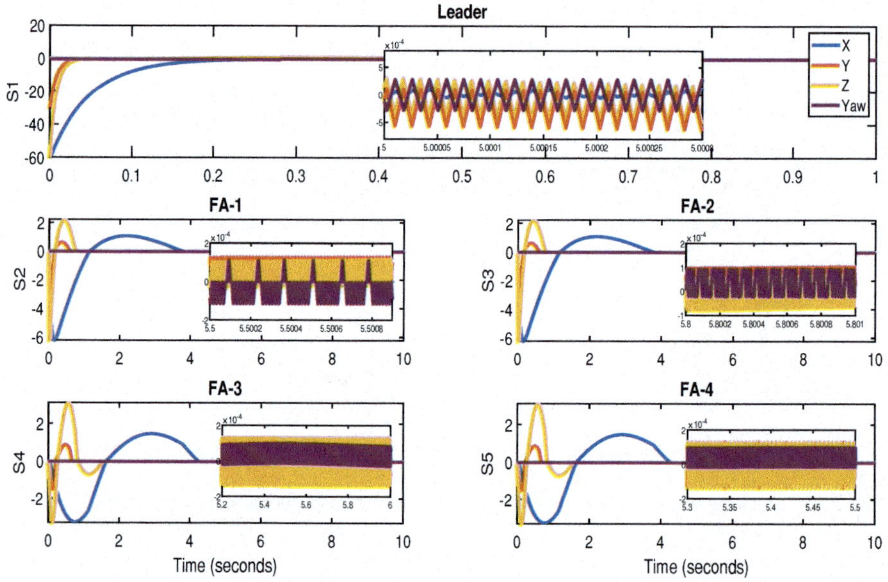

Fig. 3.13 Sliding surface of all agents using DCSM constant-proportional reaching law

3.2.8 Distributed Continuous-Time Second-Order Sliding Mode (DCSOSM) Super-Twisting (ST) Formation Protocol for Linear Model of Quadcopter Swarm

This section presents the distributed control for the leader-follower formation of the linearized model of Quadcopter swarm is designed using the DCSOSM ST protocol derived in Theorem 3.3.

Theorem 3.3 *The global homogeneous MAS given in Eq. (3.25) is said to achieve the leader-follower formation with DCSOSM ST protocol given as*

$$u(t) = [-(\bar{\Upsilon}\Gamma\bar{B}(t))^{-1}[\bar{\Upsilon}\Gamma\bar{A}(t)X(t) + \bar{\Upsilon}\Gamma(-1_5 \otimes \dot{x}_0(t) + \bar{d}_{ij})]]$$
$$-\bar{\Lambda}_1\sqrt{|\bar{\mathbb{S}}(t)|}sgn(\bar{\mathbb{S}}(t)) + \omega(t) \tag{3.56}$$

Proof Let us begin with defining the local neighborhood formation error for the MAS as

$$\delta(t) = \sum a_{ij}[[x_i(t) - x_j(t)] + \mathbb{P}_i[x_i(t) - x_0(t)] + d_{ij}] \tag{3.57}$$

where x_i and x_j are the states of the i and j agents, respectively. $\mathbb{P}_i$ is the pinning gain and d_{ij} is the distance between the agents. From Eq. (3.57), the formation error in global form can be written as

$$\mathbb{E}(t) = (((I_5 + \bar{\mathbb{D}} + \bar{\mathbb{B}})^{-1})(\bar{\mathbb{L}} + \bar{\mathbb{B}}) \otimes I_{18})(X(t) - 1_5 \otimes x_0(t) + \bar{d}_{ij}), \tag{3.58}$$

So, Eq. (3.58) is written as

$$\mathbb{E}(t) = \Gamma(X(t) - 1_5 \otimes x_0(t) + \bar{d}_{ij}) \tag{3.59}$$

where $\Gamma = (I_5 + \bar{\mathbb{D}} + \bar{\mathbb{B}})^{-1})(\bar{\mathbb{L}} + \bar{\mathbb{B}}) \otimes I_{18}))$ must be non-singular matrix.

Let us define the sliding surface for the ith agent as

$$\mathbb{S}_i(t) = \Upsilon_i\delta(t) \tag{3.60}$$

where Υ_i is the sliding gain to be evaluated from the pole-placement or LQR approach.

Now the global sliding surface can be written as

$$\bar{\mathbb{S}}(t) = \bar{\Upsilon}\mathbb{E}(t) \tag{3.61}$$

where $\bar{\Upsilon} = I_{12} \otimes \Upsilon_i$ Using Eq. (3.61), we may write

$$\bar{\mathbb{S}}(t) = \bar{\Upsilon}\Gamma(X(t) - 1_5 \otimes x_0(t) + \bar{d}_{ij}) \tag{3.62}$$

Further, advancing Eq. (3.62)

$$\dot{\bar{\mathbb{S}}}(t) = \bar{\Upsilon}\Gamma(\dot{X}(t) - 1_5 \otimes \dot{x}_0(t) + \bar{d}_{ij}) \tag{3.63}$$

Substituting Eq. (3.25) into Eq. (3.63), we may get

$$\dot{\bar{\mathbb{S}}}(t) = \bar{\Upsilon}\Gamma(\bar{A}X(t) + \bar{B}u(t) - 1_5 \otimes \dot{x}_0(t) + \bar{d}_{ij}) \tag{3.64}$$

Now, the super-twisting protocol has two parts u_s switching law, and u_{eq} equivalent law. Taking $\dot{\bar{\mathbb{S}}}(t) = 0$ and resulting u_{eq} is defined as

$$0 = \bar{\Upsilon}\Gamma(\bar{A}X(t) + \bar{B}u_{eq}(t) - 1_5 \otimes \dot{x}_0(t) + \bar{d}_{ij}) \tag{3.65}$$

After solving Eq. (3.65), we may get

$$u_{eq}(t) = [-(\bar{\Upsilon}\Gamma\bar{B})^{-1}[\bar{\Upsilon}\Gamma\bar{A}X(t) + \bar{\Upsilon}\Gamma(-1_5 \otimes \dot{x}_0(t) + \bar{d}_{ij})]] \tag{3.66}$$

Let us define the continuous-time super-twisting protocol as [9]

$$u_i sw(t) = -\Lambda_1\sqrt{|\mathbb{S}(t)|}sgn(\mathbb{S}(t)) + \omega(t)$$

$$\dot{\omega}(t) = -\Lambda_2 sgn(\mathbb{S}(t)) \tag{3.67}$$

where Λ_1, $\Lambda_2 > 0$.

The super-twisting protocol in global form can be written as

$$u_{sw}(t) = -\bar{\Lambda}_1\sqrt{|\bar{\mathbb{S}}(t)|}sgn(\bar{\mathbb{S}}(t)) + \omega(t)$$

$$\dot{\omega}(t) = -\bar{\Lambda}_2 sgn(\bar{\mathbb{S}}(t)) \tag{3.68}$$

Hence, we have the control law as

$$u(t) = u_{sw}(t) + u_{eq}(t) \tag{3.69}$$

Now, using Eq. (3.66) and Eq. (3.68), we may get the DCSOSM ST formation protocol for the linear model of Quadcopter swarm as

$$u(t) = [-(\bar{\Upsilon}\Gamma\bar{B}(t))^{-1}[\bar{\Upsilon}\Gamma\bar{A}(t)X(t) + \bar{\Upsilon}\Gamma(-1_5 \otimes \dot{x}_0(t) + \bar{d}_{ij})]] \\ -\bar{\Lambda}_1\sqrt{|\bar{\mathbb{S}}(t)|}sgn(\bar{\mathbb{S}}(t)) + \omega(t) \tag{3.70}$$

This completes the proof.

Theorem 3.4 presents the global stability of the Quadcopter swarm using DCSOSM ST formation protocol given below.

Theorem 3.4 *A global system of Quadcopter given in Eq. (3.25) is said to be asymptotically stable for super-twisting protocol, if Λ_1 and $\Lambda_2 > 0$, then trajectories of* *perturbed system goes to origin of $\bar{\mathbb{S}}(t) = 0$ in finite time $t(x)$ which is smaller than $\mathbb{T} = \frac{2\sqrt{V(x)}}{\eta}$, where x is initial state of system and η is constant depending on gains Λ_1 and Λ_2 [10].*

Proof A quadratic form of the Lyapunov function is proposed for analyzing the system's asymptotic stability as

$$\mathcal{V} = \xi^{\mathrm{T}} \mathbb{M} \xi \tag{3.71}$$

where $\xi = \begin{bmatrix} \sqrt{|\bar{\mathbb{S}}(t)|}\mathrm{sgn}(\bar{\mathbb{S}}(t)) \\ -\Lambda_2 \int \mathrm{sgn}(\bar{\mathbb{S}}(t))dt \end{bmatrix}$ and $\mathbb{M} = \frac{1}{2}\begin{bmatrix} 4\Lambda_2 + \Lambda_1^2 & -\Lambda_1 \\ -\Lambda_1 & 2 \end{bmatrix}$.

For $\Lambda_2 > 0$, a positive definite but radially unbounded above Lyapunov function $\mathcal{V}$ exists. Thus, this requires

$$\lambda_{\min}(\mathbb{M})\|\xi\|^2 \le \mathcal{V} \le \lambda_{\max}(\mathbb{M})\|\xi\|^2 \tag{3.72}$$

where $\|\xi\|^2 = |\bar{\mathbb{S}}(t)| + \left(-\Lambda_2 \int \mathrm{sgn}(\bar{\mathbb{S}}(t))dt\right)^2$ is euclidean norm of ξ. Now, a derivative of the Lyapunov function is

$$\dot{\mathcal{V}} = -\frac{1}{\sqrt{|\bar{\mathbb{S}}(t)|}}\xi^T \mathbb{N} \xi \tag{3.73}$$

where $\mathbb{N} = \frac{\Lambda_1}{2}\begin{bmatrix} 2\Lambda_2 + \Lambda_1^2 & -\Lambda_1 \\ -\Lambda_1 & 1 \end{bmatrix}$. So, the Lyapunov derivative becomes

$$\dot{\mathcal{V}} \le -\frac{1}{\sqrt{|\bar{\mathbb{S}}(t)|}}\lambda_{\min}(\mathbb{N})\|\xi\|^2 \tag{3.74}$$

Therefore, it is proved that Lyapunov derivative is symmetric negative definite for Quadcopter for $\mathbb{N} > 0$ which is done due to Λ_1 and $\Lambda_2 > 0$. After implementing a super-twisting protocol, a system is said to achieve asymptotic stability.

This completes the proof.

3.2.9 Simulation Results of DCSOSM ST Formation Protocol

In this section, simulation results of the formation of a Quadcopter swarm are discussed using the DCSOSM ST protocol. Figure 3.7 is considered for the formation of five Quadcopters. The required parameter to evaluate the global error such as the Laplacian $\mathbb{L}$, adjacency $\mathbb{A}$, degree $\mathbb{D}$, and pinning matrices $\mathbb{P}$ are also given in Eqs. (3.51–3.54).

The sliding mode gain Υ_i for the ith Quadcopter agent is designed using the LQR approach. The super-twisting constant Λ_1 and Λ_2 are chosen as 5 and 30, respectively. The output in the direction of X-, Y-, and Z-axis for all agents are shown in

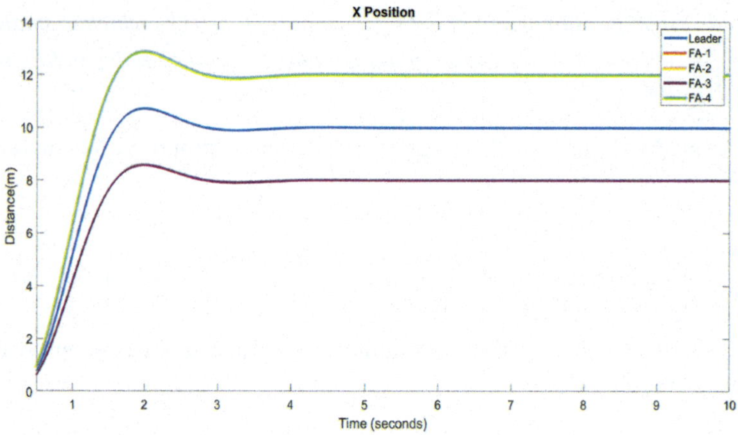

Fig. 3.14 X position of all agents using DCSOSM ST protocol

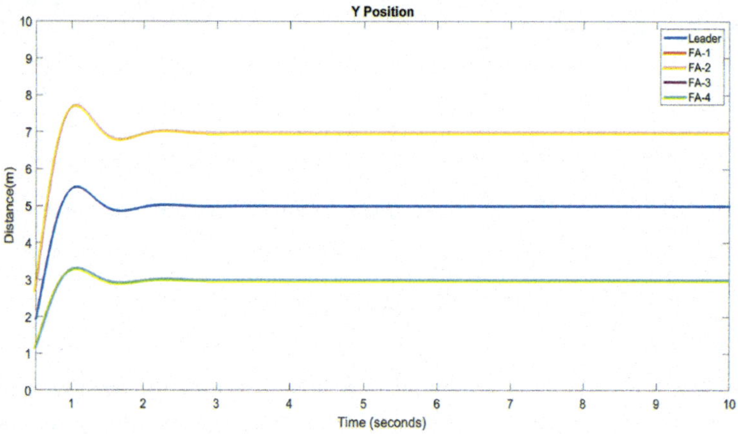

Fig. 3.15 Y position of all agents using DCSOSM ST protocol

Figs. 3.14, 3.15, and 3.16, respectively. The X-axis position of the leader is 10m from the origin. As the d_{ij} is 1m for FA-1 and FA-2 from the leader, the X position of FA-1 and FA-2 is 9m. Similarly, the X position of FA-3 and FA-4 is 8m. The d_{ij} between all agents is 1m in the Y position. And the d_{ij} is zero for the Z position, the altitude of all agents is the same at 10m.

Figure 3.17 presents the tracking error of X, Y, Z, and Yaw positions for all agents. Figure 3.18 shows the total control effort required for achieving the formation of the linear model of the Quadcopter swarm using DCSOSM ST protocol. The required control band varies between the range of -1 to $+1$. Figure 3.19 shows the sliding surface of the Quadcopter swarm. The chattering band of the sliding surface is -2×10^{-7} to $+2 \times 10^{-7}$ given in the magnified window.

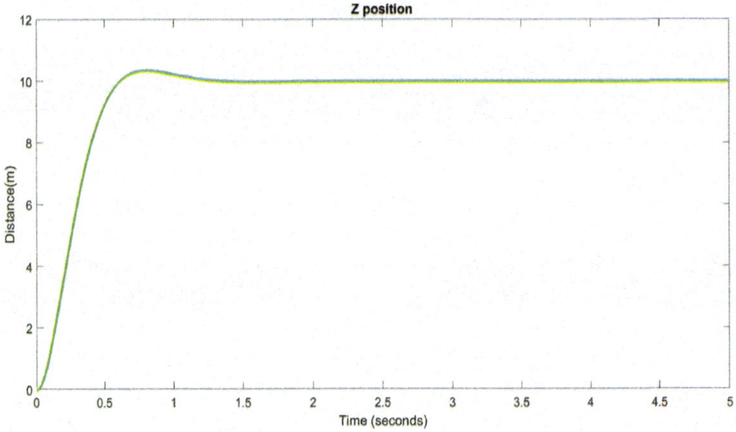

Fig. 3.16 Z position of all agents using DCSOSM ST protocol

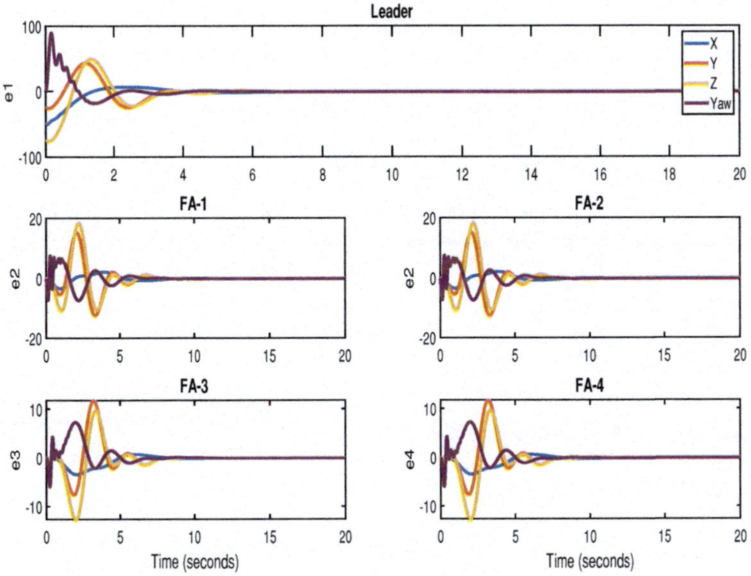

Fig. 3.17 Tracking error of all agents using DCSOSM ST protocol

Further to evaluate the performance of formation, the sliding surface variables and the control efforts of first-order DCSM protocol using constant-proportional reaching law are compared with the DCSOSM ST protocol which is quantified in Tables 3.2 and 3.3.

In Table 3.2, the chattering effect during the formation varies in the range of 10^{-4} for DCSM using constant-proportional reaching law and in the case of DCSOSM ST protocol, the chattering amplitude is reduced to the 10^{-7}. The required control

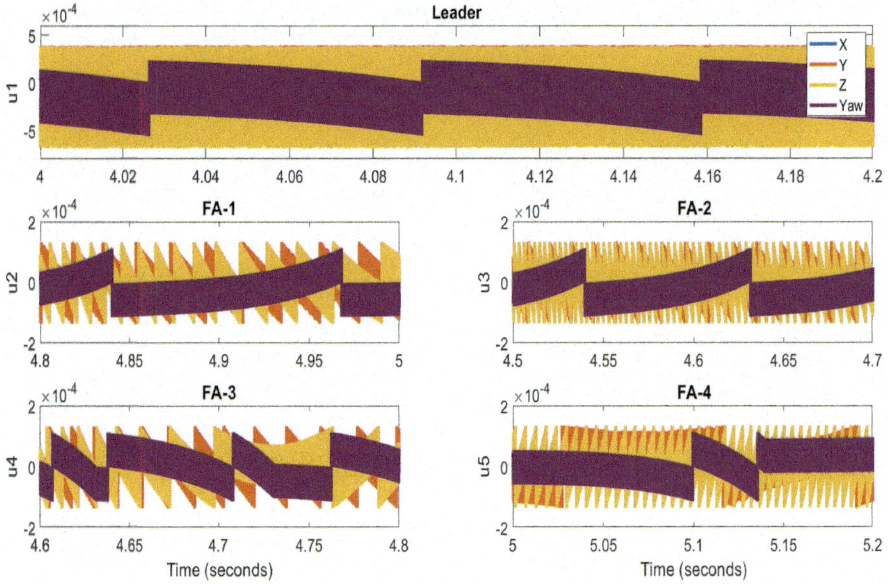

Fig. 3.18 Control efforts of all agents using DCSOSM ST protocol

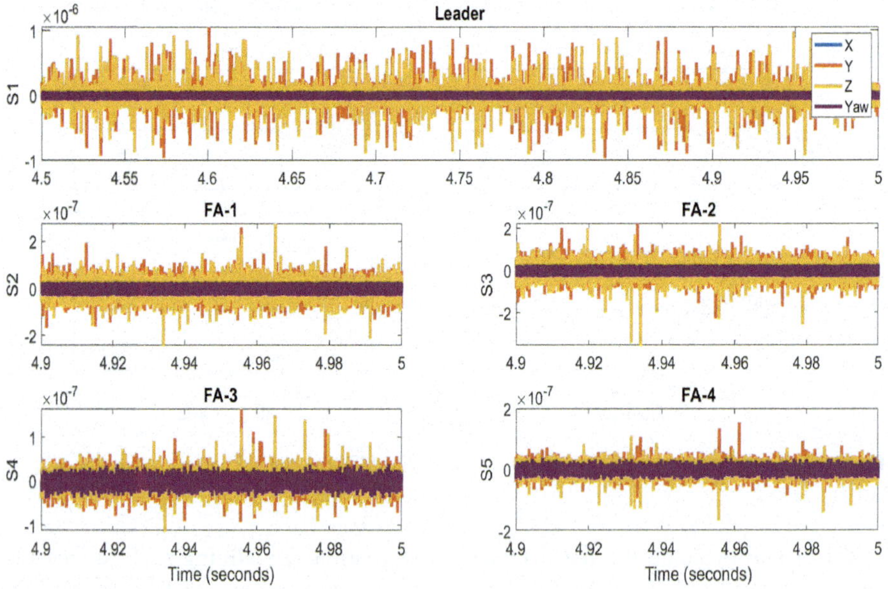

Fig. 3.19 Sliding surface of all agents using DCSOSM ST protocol

Table 3.2 Comparison study of sliding variable ($\triangle_i$)

Surface ($\bar{\mathbb{S}}$)	Constant-proportional	ST Protocol
$\mathbb{S}_1$	2.5×10^{-4}	1×10^{-6}
$\mathbb{S}_2$	1.6×10^{-4}	1.5×10^{-7}
$\mathbb{S}_3$	1×10^{-4}	1.5×10^{-7}
$\mathbb{S}_4$	1.6×10^{-4}	0.5×10^{-7}
$\mathbb{S}_5$	1.6×10^{-4}	0.5×10^{-7}

Table 3.3 Comparison study of control effort

Control effort	Constant-proportional	ST protocol
u_1	–1 to 1	–0.0005 to +0.0005
u_2	–1 to 1	–0.00015 to +0.00015
u_3	–1 to 1	–0.00015 to +0.00015
u_4	–1 to 1	–0.00015 to +0.00015
u_5	–1 to 1	–0.00015 to +0.00015

efforts to achieve the global formation of Quadcopter swarm vary in the range of -1 to 1 for the DCSM constant-proportional reaching law. And in the case of DCSOSM ST, the control efforts are reduced –0.00015 to +0.00015 given in Table 3.3.

3.3 DCSM Formation Protocol for Non-linear Model of Quadcopter Swarm

Non-linear models are crucial for more advanced control strategies, especially when the Quadcopter is subjected to varying conditions or needs to perform complex maneuvers. A linear model assumes that the relationship between input and output variables is linear. This means that the response of the system is directly proportional to changes in the inputs. But in the case of non-linear model, the relationship between inputs and outputs may not be directly proportional and may change based on operating conditions. Non-linear models can capture more intricate dynamics, such as aerodynamic effects, wind disturbances, and more.

3.3.1 Non-linear Model of Quadcopter

The mathematical model of a non-linear system can be represented in the form of $\dot{X} = f(X, U)$ by considering the $X = [x_1, x_2, x_3 \ldots x_{12}]^{\tau}$ as a state vector of the system [11].

$$\dot{X}(t) = f(X,U) = \begin{bmatrix} x_1 \\ x_2 \\ x_3 \\ x_4 \\ x_5 \\ x_6 \\ x_7 \\ x_8 \\ x_9 \\ x_{10} \\ x_{11} \\ x_{12} \end{bmatrix} = \begin{bmatrix} \dot{\phi} \\ \ddot{\phi} \\ \dot{\theta} \\ \ddot{\theta} \\ \dot{\psi} \\ \ddot{\psi} \\ \dot{z} \\ \ddot{z} \\ \dot{x} \\ \ddot{x} \\ \dot{y} \\ \ddot{y} \end{bmatrix} = \begin{bmatrix} \dot{\phi} \\ \dot{\theta}\dot{\psi}\frac{J_y - J_z}{J_x} + \frac{l}{J_x}U_2 \\ \dot{\theta} \\ \dot{\phi}\dot{\psi}\frac{J_z - J_x}{J_y} + \frac{l}{J_y}U_3 \\ \dot{\psi} \\ \dot{\theta}\dot{\phi}\frac{J_x - J_y}{J_z} + \frac{U_4}{J_z} \\ \dot{z} \\ \frac{U_1}{m}(cos\phi cos\theta) - g \\ \dot{x} \\ \frac{U_1}{m}(cos\phi cos\psi cos\theta + sin\phi sin\psi) \\ \dot{y} \\ \frac{U_1}{m}(cos\phi sin\psi cos\theta - sin\phi cos\psi) \end{bmatrix} \tag{3.75}$$

Further, the above model can be simplified as below

$$\dot{X}(t) = f(X,U) = \begin{bmatrix} x_2 \\ x_4x_6a_1 + U_2b_1 \\ x_4 \\ x_2x_6a_2 + U_3b_2 \\ x_6 \\ x_2x_4a_3 + U_4b_3 \\ x_8 \\ \frac{U_1}{m}(cosx_1cosx_3) - g \\ x_{10} \\ \frac{U_1}{m}U_x \\ x_{12} \\ \frac{U_1}{m}U_y \end{bmatrix} \tag{3.76}$$

where $U_x = cosx_1sinx_3cosx_5 + sinx_1sinx_5$, $U_y = cosx_1sinx_3sinx_5 - sinx_1cosx_5$, $a_1 = \frac{J_y - J_z}{J_x}$, $a_2 = \frac{J_z - J_x}{J_y}$, $a_3 = \frac{J_x - J_y}{J_z}$, $b_1 = \frac{l}{J_x}$, $b_2 = \frac{l}{J_y}$, and $b_3 = \frac{l}{J_z}$.

3.3.2 *Continuous-Time Primary Control for Non-linear Model of Quadcopter*

Here, the primary control strategy is designed such that the height and position of the Quadcopter can be controlled by the thrust input U_1. The remaining U_2, U_3, and U_4 are the stabilizing of roll, pitch, and yaw angle, respectively. The primary controller is designed using CSM protocol using the constant-proportional reaching law. For that, the sliding surface is designed as below,

$$\mathbb{S} = \dot{e} + \Upsilon e \tag{3.77}$$

where e = error state and Υ is the sliding mode gain calculated using LQR or pole-placement method.

1. Altitude Control: Tracking error for the altitude state is defined as

$$e_1 = Z_d - Z \tag{3.78}$$

where Z_d and Z are the desired and actual values of the altitude state.
The sliding surface for the altitude is defined as

$$\mathbb{S}_1 = \dot{e}_1 + \Upsilon e_1 \tag{3.79}$$

The derivative of Eq. (3.79) is

$$\dot{\mathbb{S}}_1 = \ddot{e}_1 + \Upsilon \dot{e}_1 \tag{3.80}$$

Furthermore,

$$\dot{\mathbb{S}}_1 = \ddot{Z}_d - \ddot{Z} + \Upsilon(\dot{Z}_d - \dot{Z}) \tag{3.81}$$

Now the constant-proportional reaching law chosen for the proposed protocol,

$$\dot{\mathbb{S}}_1 = -\lambda_1 \mathbb{S}_1 - \lambda_2 sgn(\mathbb{S}_1) \tag{3.82}$$

where λ_1 and λ_2 are the constant, where λ_1 , $\lambda_2 > 0$. Now comparing Eqs. (3.81) with (3.82) and control law $\mathbb{U}_1$ can be extracted as

$$\mathbb{U}_1 = \frac{m}{cos\phi cos\theta}[\lambda_2 sgn(\mathbb{S}_1) + \lambda_1 \mathbb{S}_1 + g + \ddot{Z}_d + \Upsilon(\ddot{Z}_d - \ddot{Z})] \tag{3.83}$$

2. Roll Control: Tracking error for the roll state is defined as

$$e_2 = \phi_d - \phi \tag{3.84}$$

where ϕ_d and ϕ are the desired and actual values of the roll state.
The sliding surface for the roll position is defined as

$$\mathbb{S}_2 = \dot{e}_2 + \Upsilon e_2 \tag{3.85}$$

The derivative of Eq. (3.85) is

$$\dot{\mathbb{S}}_2 = \ddot{e}_2 + \Upsilon \dot{e}_2 \tag{3.86}$$

Furthermore,

$$\dot{\mathbb{S}}_2 = \ddot{\phi}_d - \ddot{\phi} + \Upsilon(\dot{\phi}_d - \dot{\phi}) \tag{3.87}$$

The constant-proportional reaching law defines for the roll control as

$$\dot{\mathbb{S}}_2 = -\lambda_3\mathbb{S}_2 - \lambda_4 sgn(\mathbb{S}_2) \tag{3.88}$$

where λ_3 , $\lambda_4 > 0$. Now comparing Eqs. (3.87) with (3.88) and control law $\mathbb{U}_1$ can be extracted as

$$\mathbb{U}_2 = \frac{1}{b_1}[\lambda_4 sgn(\mathbb{S}_2) + \lambda_3\mathbb{S}_2 - a_1\ddot{\theta}\ddot{\psi} + \dddot{\phi}_d + \Upsilon(\ddot{\phi}_d - \ddot{\phi})] \tag{3.89}$$

3. Pitch Control: Tracking error for the pitch state is defined as

$$e_3 = \theta_d - \theta \tag{3.90}$$

where θ_d and θ are the desired and actual values of the pitch state.
The sliding surface for the pitch position is defined as

$$\mathbb{S}_3 = \dot{e}_3 + \Upsilon e_3 \tag{3.91}$$

The derivative of Eq. (3.91) is

$$\dot{\mathbb{S}}_3 = \ddot{e}_3 + \Upsilon\dot{e}_3 \tag{3.92}$$

Furthermore,

$$\dot{\mathbb{S}}_3 = \ddot{\theta}_d - \ddot{\theta} + \Upsilon(\dot{\theta}_d - \dot{\theta}) \tag{3.93}$$

Now define the constant-proportional reaching law for the pitch protocol as

$$\dot{\mathbb{S}}_3 = -\lambda_5\mathbb{S}_3 - \lambda_6 sgn(\mathbb{S}_3) \tag{3.94}$$

where $\lambda_5, \lambda_6 > 0$. Now comparing Eqs. (3.93) with (3.94) and control law $\mathbb{U}_3$ can be extracted as

$$\mathbb{U}_3 = \frac{1}{b_2}[\lambda_6 sgn(\mathbb{S}_3) + \lambda_5\mathbb{S}_3 - a_2\ddot{\phi}\ddot{\psi} + \dddot{\theta}_d + \Upsilon(\ddot{\theta}_d - \ddot{\theta})] \tag{3.95}$$

4. Yaw Control: Tracking error for the Yaw state is defined as

$$e_4 = \psi_d - \psi \tag{3.96}$$

where ψ_d and ψ are the desired and actual values of the yaw state.
The sliding surface for the yaw is defined as

$$\mathbb{S}_4 = \dot{e}_4 + \Upsilon e_4 \tag{3.97}$$

The derivative of Eq. (3.97) is

$$\dot{\mathbb{S}}_4 = \ddot{e}_4 + \Upsilon \dot{e}_4 \tag{3.98}$$

Furthermore,

$$\dot{\mathbb{S}}_4 = \ddot{\psi}_d - \ddot{\psi} + \Upsilon(\dot{\psi}_d - \dot{\psi}) \tag{3.99}$$

define the constant-proportional reaching law for yaw control as

$$\dot{\mathbb{S}}_4 = -\lambda_7 \mathbb{S}_4 - \lambda_8 sgn(\mathbb{S}_4) \tag{3.100}$$

where λ_7 , $\lambda_8 > 0$. Now comparing Eqs. (3.99) with (3.100) and control law $\mathbb{U}_4$ can be extracted as

$$\mathbb{U}_4 = \frac{1}{b_3}[\lambda_8 sgn(\mathbb{S}_4) + \lambda_7 S_4 - a_3 \ddot{\phi} \ddot{\theta} + \dddot{\psi}_d + c_3(\ddot{\psi}_d - \ddot{\psi})] \tag{3.101}$$

5. X Control: Tracking error for the X state is defined as

$$e_5 = X_d - X \tag{3.102}$$

where X_d and X are the desired and actual values of the X state.
The sliding surface for the X position is defined as

$$\mathbb{S}_5 = \dot{e}_5 + \Upsilon e_5 \tag{3.103}$$

The derivative of Eq. (3.103) is

$$\dot{\mathbb{S}}_5 = \ddot{e}_5 + \Upsilon \dot{e}_5 \tag{3.104}$$

Furthermore,

$$\dot{\mathbb{S}}_5 = \ddot{X}_d - \ddot{X} + \Upsilon(\dot{X}_d - \dot{X}) \tag{3.105}$$

The constant-proportional reaching law for the X position is given as

$$\dot{\mathbb{S}}_5 = -\lambda_9 \mathbb{S}_5 - \lambda_{10} sgn(\mathbb{S}_5) \tag{3.106}$$

where λ_9 , $\lambda_{10} > 0$. Now comparing Eqs. (3.105) with (3.106) and control law $\mathbb{U}_x$ can be extracted as

$$\mathbb{U}_x = \frac{m}{U_1}[\lambda_9 sgn(\mathbb{S}_5) + \lambda_{10} S_5 + \dddot{X}_d + c_5(\ddot{X}_d - \ddot{x})] \tag{3.107}$$

6. Y Control: Tracking error for the Y state is defined as

$$e_6 = Y_d - Y \tag{3.108}$$

where Y_d and Y are the desired and actual values of the Y state.
The sliding surface for the Y position is defined as

$$\mathbb{S}_6 = \dot{e}_6 + \Upsilon e_6 \tag{3.109}$$

The derivative of Eq. (3.109) is

$$\dot{\mathbb{S}}_6 = \ddot{e}_6 + \Upsilon \dot{e}_6 \tag{3.110}$$

Furthermore,

$$\dot{\mathbb{S}}_6 = \ddot{Y}_d - \ddot{Y} + \Upsilon(\dot{Y}_d - \dot{Y}) \tag{3.111}$$

The constant-proportional reaching law for the Y position is given as

$$\dot{\mathbb{S}}_6 = -\lambda_{11}\mathbb{S}_6 - \lambda_{12} sgn(\mathbb{S}_6) \tag{3.112}$$

where λ_{11} , $\lambda_{12} > 0$. Now comparing Eqs. (3.111) with (3.112) and control law $\mathbb{U}_y$ can be extracted as

$$\mathbb{U}_y = \frac{m}{U_1}[\lambda_{11} sgn(\mathbb{S}_6) + \lambda_{12} S_6 + \dddot{y_d} + c_6(\ddot{y}_d - \ddot{y})] \tag{3.113}$$

Here U_1, U_2, U_3, and U_4 are responsible for generating the Z, roll, pitch, and yaw values.

3.3.3 *Simulation Results of Continuous-Time Primary Control for Non-linear Model of Quadcopter*

In this section, the simulation results for non-linear model of the Quadcopter system are discussed with sliding mode control using the constant-proportional reaching law. The value of all the reaching law constants is chosen as 3 and the sliding mode gain Υ chosen as 2. Here, one finite-time trajectory is given to the single Quadcopter for X, Y, and Z positions shown in Fig. 3.20. The required control efforts to achieve the formation are shown in Fig. 3.21. The sliding surface and the tracking error for the Quadcopter are given in Figs. 3.22 and 3.23.

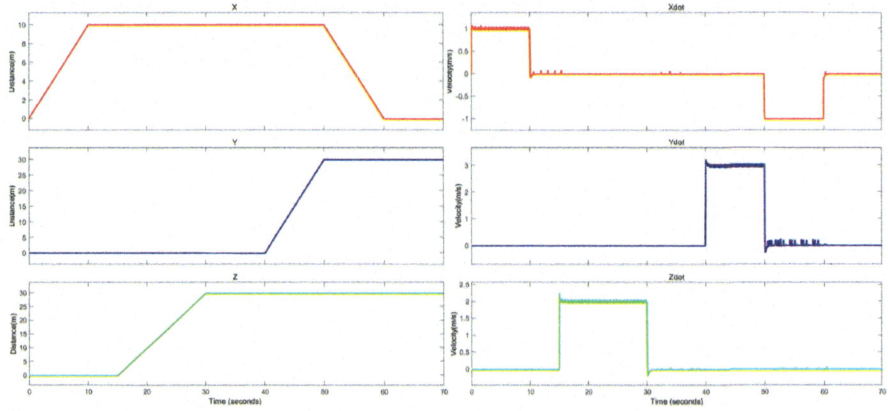

Fig. 3.20 Transitional motion of non-linear model of Quadcopter

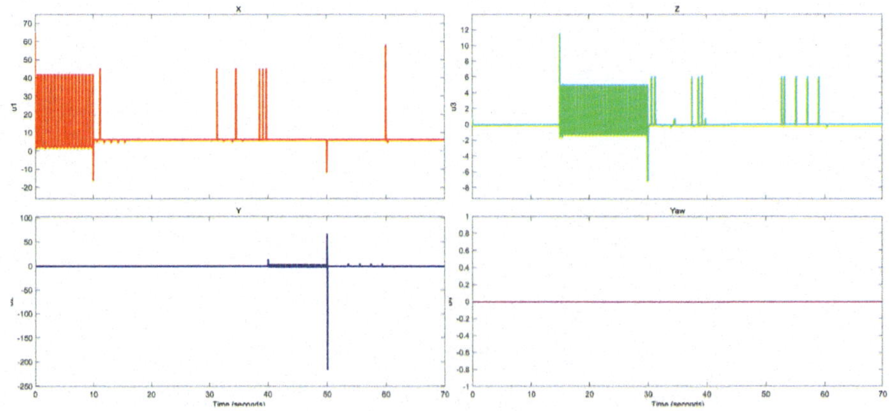

Fig. 3.21 Control efforts of non-linear model of Quadcopter

3.3.4 DCSOSM ST Formation Protocol for Non-linear Model of Quadcopter Swarm

This Sect. 3.3.4 provides the detailed information about the DCSOSM formation protocol for non-linear model of Quadcopter swarm. With the help of graph theory, the agents have their neighborhood agents information which is shared by the undirected graph topology. To design the distributed continuous-time sliding mode formation protocol, the pattern of the five Quadcopters as shown in Fig. 3.24 is considered.

As per the formation pattern in Fig. 3.25, the position of the leader agent is considered as the (0, 0, 0) for the (X, Y, Z) position. The immediate follower-1 (FA-1) and follower-2 (FA-2) have the distance d_{ij} with the leader agent in (X, Y, Z) position

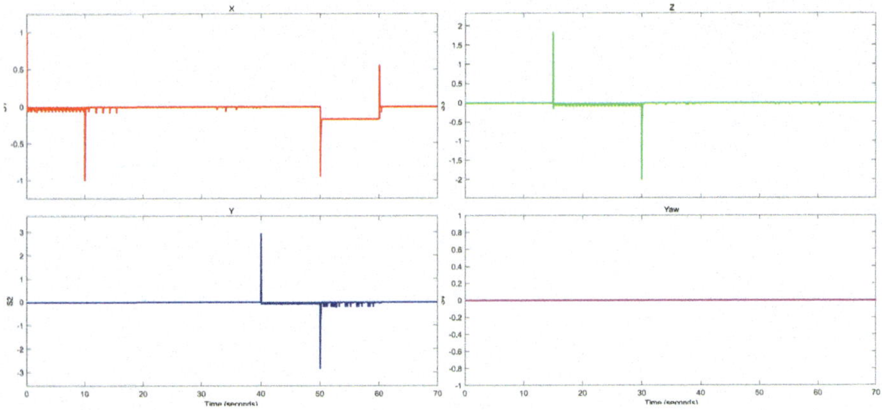

Fig. 3.22 Sliding surface of non-linear model of Quadcopter

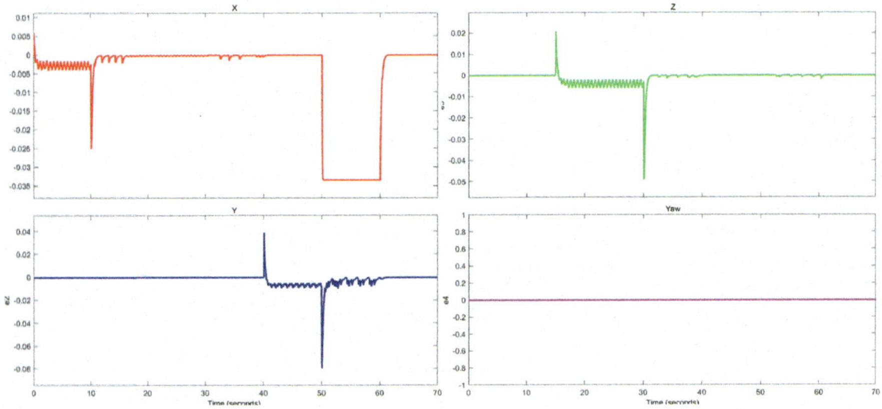

Fig. 3.23 Tracking error of non-linear model of Quadcopter

as (1, 1, 0) and (1, –1, 0). Then the next follower-3 (FA-3) and follower-4 (FA-4) have the distance d_{ij} with leader agent in (X, Y, Z) position as (2, 2, 0) and (2, –2, 0). The altitude of all the Quadcopters will remain the same or it can be presented as $d_{ij} = 0$ for all Quadcopters in the Z position.

The next theorem presents the DCSOSM ST protocol for the formation of Quadcopter swarm.

Theorem 3.5 *The global homogeneous MAS given in Eq. (3.115) is said to achieve the leader-follower formation with the DCSOSM ST protocol given below*

Consider the mathematical model for the non-linear Quadcopter given in Sect. 3.3.5 which can be represented for nth order as

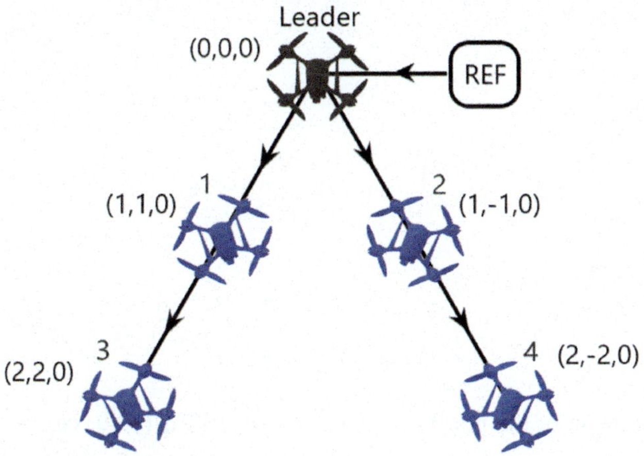

Fig. 3.24 Communication Topology

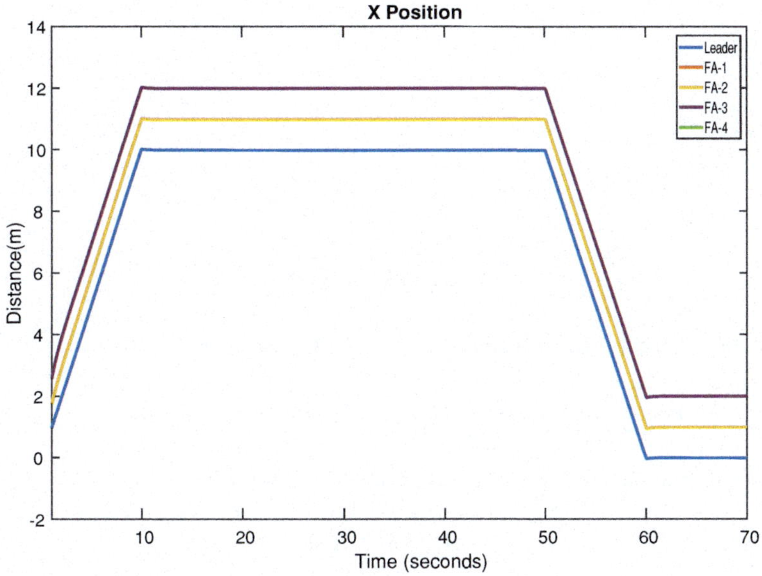

Fig. 3.25 X trajectory for all agents using DCSOSM ST protocol

$$\dot{x}_1 = x_2$$

$$\dot{x}_2 = x_3$$

$$\vdots$$

$$\dot{x}_{n-1} = x_n$$

$$\dot{x}_n = f(x) + g(x)u \tag{3.114}$$

The global form of the non-linear model given in Eq. (3.89) can be written as

$$\dot{X}_1 = X_2$$

$$\dot{X}_2 = X_3$$

$$\vdots$$

$$\dot{X}_{n-1} = X_n$$

$$\dot{X}_n = f(X) + g(X)u \tag{3.115}$$

The local neighborhood formation error for the MAS as

$$\delta(t) = \sum a_{ij}[[x_i(t) - x_j(t)] + \mathbb{P}_i[x_i(t) - x_0(t)] + d_{ij}] \tag{3.116}$$

Let us define the sliding surface for the ith agent as

$$\mathbb{S}_i = \dot{\delta}_i + \Upsilon_i \delta_i \tag{3.117}$$

where ${\Upsilon_i}^T$ is the sliding mode gain for the ith agent.

Now, a super-twisting protocol is used for the formation of Quadcopter swarm. A switching control action is developed using a DCSOSM ST protocol by the Lyapunov function approach which shows the asymptotic stability of a system. A switching control law u_{isw} for ith is designed as

$$u_{isw}(t) = -\Lambda_1 \sqrt{|\mathbb{S}_i|} sgn(\mathbb{S}_i) + \omega(t)$$

$$\dot{\omega}(t) = -\Lambda_2 sgn(\mathbb{S}_i) \tag{3.118}$$

where $\Lambda_1, \Lambda_2 > 0$.

A switching control action is used to enforce system trajectories of Quadcopters on a sliding surface. Once they reach the sliding surface, they are free from system parameters and are dependent only on sliding surface parameters.

This completes the proof.

The stability of the DCSOSM ST formation protocol for the Quadcopter swarm is given in Sect. 3.2.8 Theorem 3.4.

3.3.5 Simulation Results of Non-linear model of Quadcopter using DCSOSM ST formation Protocol

In this section, simulation results of the formation of non-linear model of Quadcopter swarm are discussed using the DCSOSM ST protocol. All agents shared their information to the neighborhood agent as per the formation pattern. The summation of these information is known as the local error which is given in Eq. (3.91). The sliding mode gain Υ_i for the ith Quadcopter agent is chosen 2. The super-twisting constant Λ_1 and Λ_2 are chosen as 2 and 5, respectively.

Here, one finite-time trajectory is given to the leader for X, Y, and Z positions shown in Figs. 3.25, 3.26, and 3.27, respectively. As per the formation pattern, FA-1 and FA-2 have the distance d_{ij} of +1m from the leader agent in X position. FA-3 and FA-4 have the d_{ij} is +2m from the leader agent in X position. for FA-1 and FA-2 from the leader, the X position of FA-1 and FA-2 is 9m. Similarly for Y position, d_{ij} is 1m between all agents. And the d_{ij} is zero for the Z position, the altitude of

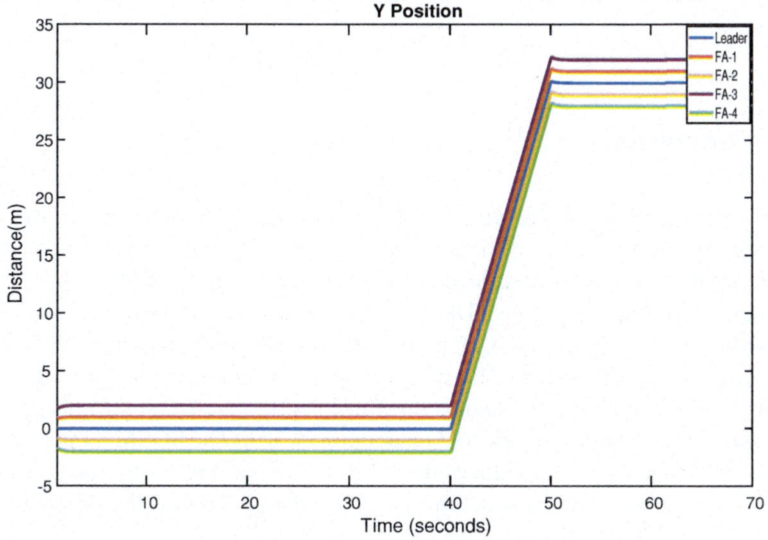

Fig. 3.26 Y trajectory for all agents using DCSOSM ST protocol

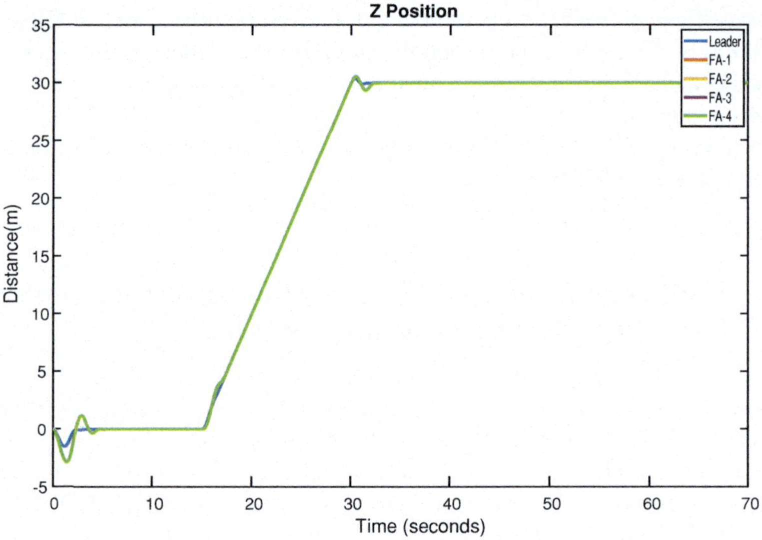

Fig. 3.27 Z trajectory for all agents using DCSOSM ST protocol

all agents is the same throughout the flight. Figure 3.28 is the tracking error for all five Quadcopters. Figure 3.29 shows the required control effort during the trajectory for all Quadcopters. Sliding variables of all agents are shown in Fig. 3.30. Hence to check the robustness of the formation of five Quadcopters, the external disturbance is given to the X and Y trajectory shown in Fig. 3.31. Even if there is a variation in mass while using the payload, still the DCSOSM ST formation protocol will provide better results.

3.4 Conclusion

This chapter discusses an analytical evaluation of the continuous-time sliding mode protocols for the formation of linear and non-linear models of Quadcopter swarm. The concept of MAS known as leader-follower formation is configured with a fixed and undirected graph topology. The first-order constant-proportional reaching law and second-order super-twisting type of distributed continuous-time sliding mode (DCSM) protocol are designed for a linear model of Quadcopter swarm formation. In order to test the robustness against the non-linearity, the DCSOSM ST protocol is designed for the non-linear model of Quadcopter swarm. The SOSM ST algorithm not only ensures robustness but reduces the chattering effect and faster the convergence time. The Lyapunov stability function is used for the global stability of the formation

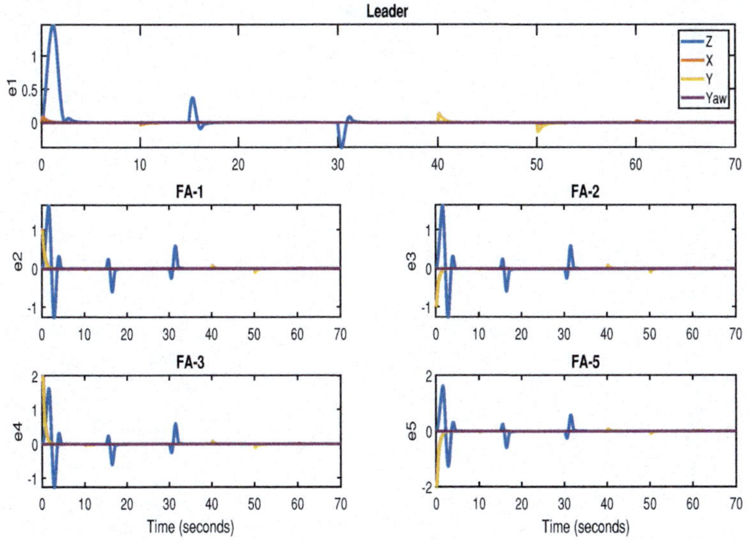

Fig. 3.28 Control efforts for all agents using DCSOSM ST protocol

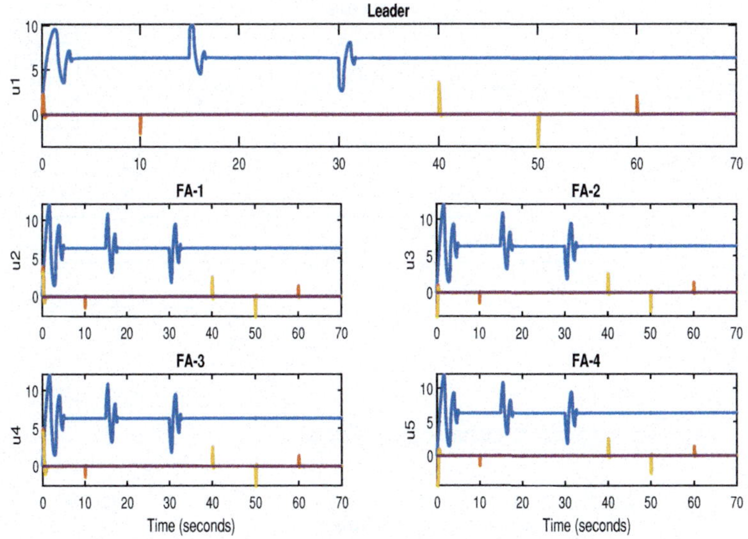

Fig. 3.29 Control efforts for all agents using DCSOSM ST protocol

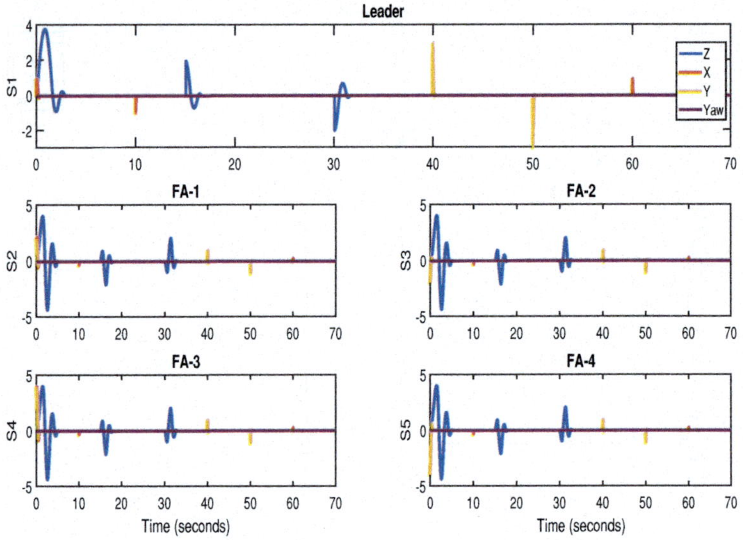

Fig. 3.30 Sliding surface for all agents using DCSOSM ST protocol

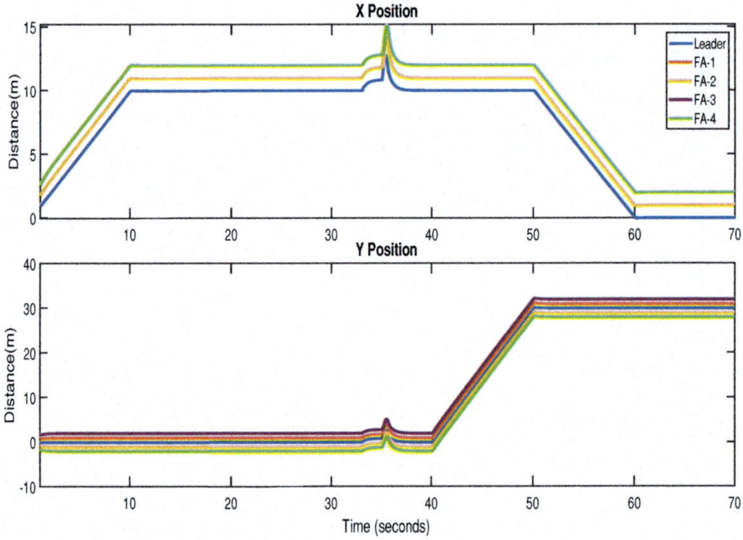

Fig. 3.31 X & Y position for all agents under the disturbance

of the swarm. From the simulation results, it is inferred that the DCSM formation protocol brings all Quadcopters in the given pattern in a finite time. The robustness of the protocol under external disturbance and parameter variation is also tested.

References

1. Wang, P., Man, Z., Cao, Z., Zheng, J., Zhao, Y.: Dynamics modelling and linear control of quadcopter. In: 2016 International Conference on Advanced Mechatronic Systems (ICAMechS), Melbourne, VIC, Australia, 2016, pp. 498–503 (2016). https://doi.org/10.1109/ICAMechS.2016.7813499
2. Ahmad, F., Kumar, P., Bhandari, A., Patil, P.P.: Simulation of the quadcopter dynamics with LQR based control. In: Materials Today: Proceedings, vol. 24, Part 2, pp. 326–332 (2020). ISSN 2214-7853. https://doi.org/10.1016/j.matpr.2020.04.282
3. Patel, K., Mehta, A.: Discrete-time sliding mode protocols for discrete multi-agent system, vol. 303, pp. 1–121. Springer (2021). https://doi.org/10.1007/978-981-15-6311-9
4. Patel, K., Mehta, A.: Discrete-time sliding mode protocols for leader-following consensus of discrete multi-agent system with switching graph topology. Eur. J. Control (2020). ISSN 0947-3580. https://doi.org/10.1016/j.ejcon.2019.06.011
5. Patel, K., Mehta, A.: Discrete Higher order sliding mode protocol for leader-following consensus of heterogeneous discrete multi-agent system. In: Mehta, A., Rawat, A., Chauhan, P. (eds.) Advances in Control Systems and its Infrastructure. Lecture Notes in Electrical Engineering, vol. 604. Springer, Singapore (2020). https://doi.org/10.1007/978-981-15-0226-2_1
6. De Queiroz, M., Cai, X., Feemster, M.: Formation control of Multi-agent systems: a graph rigidity approach. Wiley (2019). https://doi.org/10.1002/9781118887455
7. Milosavljevic, C.: General condition for the existence of a quasi-sliding mode in the switching hyperplane in discrete variable structure system. Autom. Remote Control **46**, 307–314 (1985)
8. Corless, M., Manela, J.: Control of uncertain discrete-time systems. In: 1986 American Control Conference, pp. 515–520 (1986)
9. Kamal, S., Chalanga, A., Moreno, J.A., Fridman, L., Bandyopadhyay, B.: Higher order super-twisting algorithm. In: 13th International Workshop on Variable Structure Systems (VSS), vol. 2014, pp. 1–5. Nantes, France (2014). https://doi.org/10.1109/VSS.2014.6881129
10. Moreno, J.A., Osorio, M.: Strict Lyapunov functions for the super-twisting algorithm. IEEE Trans. Autom. Control **57**(4):1035–1040 (2012). https://doi.org/10.1109/TAC.2012.2186179
11. Tripathi, V.K., Behera, L., Verma, N.: Design of sliding mode and backstepping controllers for a quadcopter. In: 39th National Systems Conference (NSC), vol. 2015, pp. 1–6. Greater Noida, India (2015). https://doi.org/10.1109/NATSYS.2015.7489097

Chapter 4
Discrete-Time Sliding Mode Protocol for Formation of Quadcopter Swarm

4.1 System Description and Mathematical Modeling of Quadcopter

This Sect. 4.1 gives the information about the system description and the mathematical modeling of the Quadcopter provided by the Quanser PVT. LTD [1]. This model consists of 18 different states of the Quadcopter. Figure 3.1 shows the setup for the formation of multiple Quadcopters. The Quadcopter has four propellers connected to four BLDC motors. The Quadcopter's electronics are housed in a central frame, which also contains the battery and the flight controller and it receives inputs from various sensors to determine the Quadcopter's orientation and position. These sensors typically include:

- Accelerometers: measure the acceleration of the Quadcopter in three axes.
- Gyroscopes: measure the angular velocity of the Quadcopter in three axes.
- Magnetometers: measure the orientation of the Quadcopter power rate of the Earth's magnetic field.
- Barometers: measure the atmospheric pressure to estimate the altitude of the Quadcopter.

To accomplish the goal, let us derive the state space model of a Quadcopter using the kinematics and dynamics of a Quadcopter. According to Newton's second law of motion, the applied torque results in an angular acceleration given by

$$\ddot{\phi} = \frac{1}{I_\phi}\tau_\phi \tag{4.1}$$

$$\ddot{\theta} = \frac{1}{I_\theta}\tau_\theta \tag{4.2}$$

A. Mehta and A. Modi, *Robust Sliding Mode Protocols for Formation of Quadcopter Swarm*, Studies in Systems, Decision and Control 521,
https://doi.org/10.1007/978-981-99-9726-8_4

$$\ddot{\psi} = \frac{1}{I_\psi}\tau_\psi \tag{4.3}$$

where the angular accelerations of roll, pitch, and yaw for the specified model are denoted by $\ddot{\phi}$, $\ddot{\theta}$, and $\ddot{\psi}$, respectively. The roll, pitch, and yaw moments of inertia are denoted as I_ϕ, I_θ, and I_ψ, respectively. The equations of angular motion power rate the commanded torques become as

$$\dddot{\phi} = -\frac{\ddot{\phi}}{\tau_k} + \frac{\tau_\phi}{I_\phi \tau_k} \tag{4.4}$$

$$\dddot{\theta} = -\frac{\ddot{\theta}}{\tau_k} + \frac{\tau_\theta}{I_\theta \tau_k} \tag{4.5}$$

$$\dddot{\psi} = -\frac{\ddot{\psi}}{\tau_k} + \frac{\tau_\psi}{I_\psi \tau_k} \tag{4.6}$$

where τ_k is the time constant and the corresponding angular jerks are $\dddot{\phi}$, $\dddot{\theta}$, and $\dddot{\psi}$. The linear motion will likewise be segmented into X, Y, and Z-axis locations in a manner similar to the angular motion scenario. Their dynamics are as

$$\ddot{X} = \frac{T_h}{m} s\theta c\phi \tag{4.7}$$

$$\ddot{Y} = -\frac{T_h}{m} s\phi \tag{4.8}$$

$$\ddot{Z} = \frac{T_h}{m} c\theta c\phi - g \tag{4.9}$$

where T_h, m, and g are the thrust produced by the Quadcopter's motors, mass, and gravitational force, respectively. The linear acceleration in the X, Y, and Z - axis are denoted as $\ddot{X}$, $\ddot{Y}$, and $\ddot{Z}$. The equations of linear motion power rate the commanded torques become as

$$\dddot{X} = -\frac{\ddot{X}}{\tau_k} + \frac{T_h \theta}{m \tau_k} \tag{4.10}$$

$$\dddot{Y} = -\frac{\ddot{Y}}{\tau_k} + \frac{T_h \phi}{m \tau_k} \tag{4.11}$$

$$\dddot{Z} = -\frac{\ddot{Z}}{\tau_k} + \frac{T_h}{m \tau_k} - \frac{g}{\tau_k} \tag{4.12}$$

where $\dddot{X}$, $\dddot{Y}$, and $\dddot{Z}$ are the linear jerks in X-, Y-, and Z-axis, respectively.

Table 4.1 System parameters of Quadcopter

Parameters	
θ	Pitch angle
ϕ	Roll angle
ψ	Yaw angle
I_ϕ	Roll moment of inertia = 10^{-2} kgm^2
I_θ	Pitch moment of inertia = 8.2×10^{-2} kgm^2
I_ψ	Yaw moment of inertia = 1.48×10^{-2} kgm^2
τ_ϕ	Roll torque
τ_θ	Pitch torque
τ_ψ	Yaw torque
T_h	Total thrust
τ_k	Time constant = 40 ms
m	Total mass = 1.121 kg
g	Gravitational constant = 9.81 ms^{-2}

Table 4.2 Motor and Propellers specifications

Specifications	
K_v	2100 RPM/V
Stator Diameter/thickness	22/6 mm
Stator slots/magnet poles	12/14
Max cont. current	25 A
Time constant	40 ms
Diameter	6 inches
Pitch	4.5 inches
Material	Poly-carbonate

Tables 4.1, 4.2 and 4.3 give the complete details of the Quadcopter, motors, propellers, and the onboard battery which will provide the power.

The identical state space model for the linear time-invariant system in continuous-time is given below

$$\dot{X}(t) = A(t)x(t) + B(t)u(t)$$

$$Y(t) = C(t)x(t) \tag{4.13}$$

Table 4.3 Specification of Battery

Battery characteristics		
1	Cells	3S
2	Capacity	3300 mAh
3	Cont. discharge rating	50 C
4	Max voltage per cell	4.2 V
5	Normal voltage per cell	3.7 V
6	Min voltage per cell	3.5 V
7	Battery weight	267 gm

where A(t) and B(t) are the plant matrix and input matrix, respectively. x(t) is the state vector and u(t) is an input vector and C(t) is 18 $\times$ 18 identity matrix. This model contains 18 different states of angular and linear motion shown in Eq. (4.14).

$$x(t) = \begin{bmatrix} x_1 \\ x_2 \\ x_3 \\ x_4 \\ x_5 \\ x_6 \\ x_7 \\ x_8 \\ x_9 \\ x_{10} \\ x_{11} \\ x_{12} \\ x_{13} \\ x_{14} \\ x_{15} \\ x_{16} \\ x_{17} \\ x_{18} \end{bmatrix} = \begin{bmatrix} \phi \\ \theta \\ \psi \\ \dot{\phi} \\ \dot{\theta} \\ \dot{\psi} \\ \ddot{\phi} \\ \ddot{\theta} \\ \ddot{\psi} \\ x \\ y \\ z \\ \dot{x} \\ \dot{y} \\ \dot{z} \\ \ddot{x} \\ \ddot{y} \\ \ddot{z} \end{bmatrix} = \begin{bmatrix} Roll\ position \\ Pitch\ position \\ Yaw\ position \\ Velocity\ of\ roll \\ Velocity\ of\ pitch \\ Velocity\ of\ yaw \\ Acceleration\ of\ roll \\ Acceleration\ of\ pitch \\ Acceleration\ of\ yaw \\ Linear\ position\ in\ X-axis \\ Linear\ position\ in\ Y-axis \\ Linear\ position\ in\ Z-axis \\ Linear\ velocity\ in\ X-axis \\ Linear\ velocity\ in\ Y-axis \\ Linear\ velocity\ in\ Z-axis \\ Linear\ acceleration\ in\ X-axis \\ Linear\ acceleration\ in\ Y-axis \\ Linear\ acceleration\ in\ Z-axis \end{bmatrix} \tag{4.14}$$

From the dynamical equations, the state space model of the Quadcopter in the continuous-time domain is derived as given Eqs. (4.15) and (4.16).

$$
A(t)=\begin{bmatrix}
0 & 0 & 0 & 1 & 0 & 0 & 0 & 0 & 0 & 0 & 0 & 0 & 0 & 0 & 0 & 0 & 0 & 0\\
0 & 0 & 0 & 0 & 1 & 0 & 0 & 0 & 0 & 0 & 0 & 0 & 0 & 0 & 0 & 0 & 0 & 0\\
0 & 0 & 0 & 0 & 0 & 1 & 0 & 0 & 0 & 0 & 0 & 0 & 0 & 0 & 0 & 0 & 0 & 0\\
0 & 0 & 0 & 0 & 0 & 0 & 1 & 0 & 0 & 0 & 0 & 0 & 0 & 0 & 0 & 0 & 0 & 0\\
0 & 0 & 0 & 0 & 0 & 0 & 0 & 1 & 0 & 0 & 0 & 0 & 0 & 0 & 0 & 0 & 0 & 0\\
0 & 0 & 0 & 0 & 0 & 0 & 0 & 0 & 1 & 0 & 0 & 0 & 0 & 0 & 0 & 0 & 0 & 0\\
0 & 0 & 0 & 0 & 0 & 0 & -\frac{1}{\tau_k} & 0 & 0 & 0 & 0 & 0 & 0 & 0 & 0 & 0 & 0 & 0\\
0 & 0 & 0 & 0 & 0 & 0 & 0 & -\frac{1}{\tau_k} & 0 & 0 & 0 & 0 & 0 & 0 & 0 & 0 & 0 & 0\\
0 & 0 & 0 & 0 & 0 & 0 & 0 & 0 & -\frac{1}{\tau_k} & 0 & 0 & 0 & 0 & 0 & 0 & 0 & 0 & 0\\
0 & 0 & 0 & 0 & 0 & 0 & 0 & 0 & 0 & 0 & 0 & 0 & 1 & 0 & 0 & 0 & 0 & 0\\
0 & 0 & 0 & 0 & 0 & 0 & 0 & 0 & 0 & 0 & 0 & 0 & 0 & 1 & 0 & 0 & 0 & 0\\
0 & 0 & 0 & 0 & 0 & 0 & 0 & 0 & 0 & 0 & 0 & 0 & 0 & 0 & 1 & 0 & 0 & 0\\
0 & 0 & 0 & 0 & 0 & 0 & 0 & 0 & 0 & 0 & 0 & 0 & 0 & 0 & 0 & 0 & 0 & 0\\
0 & 0 & 0 & 0 & 0 & 0 & 0 & 0 & 0 & 0 & 0 & 0 & 0 & 0 & 0 & 0 & 0 & 0\\
0 & 0 & 0 & 0 & 0 & 0 & 0 & 0 & 0 & 0 & 0 & 0 & 0 & 0 & 0 & 0 & 0 & 0\\
0 & \frac{g}{\tau_k} & 0 & 0 & 0 & 0 & 0 & 0 & 0 & 0 & 0 & 0 & 0 & 0 & 0 & -\frac{1}{\tau_k} & 0 & 0\\
-\frac{g}{\tau_k} & 0 & 0 & 0 & 0 & 0 & 0 & 0 & 0 & 0 & 0 & 0 & 0 & 0 & 0 & 0 & -\frac{1}{\tau_k} & 0\\
0 & 0 & 0 & 0 & 0 & 0 & 0 & 0 & 0 & 0 & 0 & 0 & 0 & 0 & 0 & 0 & 0 & -\frac{1}{\tau_k}
\end{bmatrix}; \tag{4.15}
$$

$$
B(t)=\begin{bmatrix}
0 & 0 & 0 & 0\\
0 & 0 & 0 & 0\\
0 & 0 & 0 & 0\\
0 & 0 & 0 & 0\\
0 & 0 & 0 & 0\\
0 & 0 & 0 & 0\\
0 & \frac{1}{I_\phi \tau_k} & 0 & 0\\
0 & 0 & \frac{1}{I_\theta \tau_k} & 0\\
0 & 0 & 0 & \frac{1}{I_\psi \tau_k}\\
0 & 0 & 0 & 0\\
0 & 0 & 0 & 0\\
0 & 0 & 0 & 0\\
0 & 0 & 0 & 0\\
0 & 0 & 0 & 0\\
0 & 0 & 0 & 0\\
0 & 0 & 0 & 0\\
0 & 0 & 0 & 0\\
\frac{1}{m\tau_k} & 0 & 0 & 0
\end{bmatrix} \tag{4.16}
$$

$$
A(t) = \begin{bmatrix}
0&0&0&1&0&0&0&0&0&0&0&0&0&0&0&0&0&0\\
0&0&0&0&1&0&0&0&0&0&0&0&0&0&0&0&0&0\\
0&0&0&0&0&1&0&0&0&0&0&0&0&0&0&0&0&0\\
0&0&0&0&0&0&1&0&0&0&0&0&0&0&0&0&0&0\\
0&0&0&0&0&0&0&1&0&0&0&0&0&0&0&0&0&0\\
0&0&0&0&0&0&0&0&1&0&0&0&0&0&0&0&0&0\\
0&0&0&0&0&0&-25&0&0&0&0&0&0&0&0&0&0&0\\
0&0&0&0&0&0&0&-25&0&0&0&0&0&0&0&0&0&0\\
0&0&0&0&0&0&0&0&-25&0&0&0&0&0&0&0&0&0\\
0&0&0&0&0&0&0&0&0&0&0&0&1&0&0&0&0&0\\
0&0&0&0&0&0&0&0&0&0&0&0&0&1&0&0&0&0\\
0&0&0&0&0&0&0&0&0&0&0&0&0&0&1&0&0&0\\
0&0&0&0&0&0&0&0&0&0&0&0&0&0&0&0&0&0\\
0&0&0&0&0&0&0&0&0&0&0&0&0&0&0&0&0&0\\
0&0&0&0&0&0&0&0&0&0&0&0&0&0&0&0&0&0\\
0&245.25&0&0&0&0&0&0&0&0&0&0&0&0&0&-25&0&0\\
-245.25&0&0&0&0&0&0&0&0&0&0&0&0&0&0&0&-25&0\\
0&0&0&0&0&0&0&0&0&0&0&0&0&0&0&0&0&-25
\end{bmatrix}; \tag{4.17}
$$

$$
B(t) = \begin{bmatrix}
0&0&0&0\\
0&0&0&0\\
0&0&0&0\\
0&0&0&0\\
0&0&0&0\\
0&0&0&0\\
0&2500&0&0\\
0&0&304.8&0\\
0&0&0&1689.19\\
0&0&0&0\\
0&0&0&0\\
0&0&0&0\\
0&0&0&0\\
0&0&0&0\\
0&0&0&0\\
0&0&0&0\\
0&0&0&0\\
22.30&0&0&0
\end{bmatrix} \tag{4.18}
$$

Nowadays in the application of Quadcopter swarm, digital controllers and processors are being used for communication. However, it is always challenging to implement the continuous-time sliding mode protocol for formation using a digital processor due to high-frequency control efforts for maintaining the formation among the agents. Further, the system is discretized at $\tau = 0.001$ sampling time to obtain the discrete-time model given in Eqs. (4.19) and (4.20).

$$E(k) = \begin{bmatrix} 1 & 0 & 0 & 0.001 & 0 & 0 & 0 & 0 & 0 & 0 & 0 & 0 & 0 & 0 & 0 & 0 & 0 & 0 \\ 0 & 1 & 0 & 0 & 0 & 0.001 & 0 & 0 & 0 & 0 & 0 & 0 & 0 & 0 & 0 & 0 & 0 & 0 \\ 0 & 0 & 1 & 0 & 0 & 0 & 0 & 0.001 & 0 & 0 & 0 & 0 & 0 & 0 & 0 & 0 & 0 & 0 \\ 0 & 0 & 0 & 1 & 0 & 0 & 0 & 0 & 0 & 0 & 0 & 0 & 0 & 0 & 0 & 0 & 0 & 0 \\ 0 & 0 & 0 & 0 & 1 & 0 & 0 & 0 & 0 & 0 & 0 & 0 & 0 & 0 & 0 & 0 & 0 & 0 \\ 0 & 0 & 0 & 0 & 0 & 1 & 0 & 0 & 0 & 0 & 0 & 0 & 0 & 0 & 0 & 0 & 0 & 0 \\ 0 & 0 & 0 & 0 & 0 & 0 & 1 & 0 & 0 & 0 & 0 & 0 & 0 & 0 & 0 & 0 & 0 & 0 \\ 0 & 0 & 0 & 0 & 0 & 0 & 0 & 1 & 0 & 0 & 0 & 0 & 0 & 0 & 0 & 0 & 0 & 0 \\ 0 & 0 & 0 & 0 & 0 & 0 & 0 & 0 & 0.9753 & 0 & 0 & 0 & 0 & 0 & 0 & 0 & 0 & 0 \\ 0 & 0 & 0 & 0 & 0 & 0 & 0 & 0 & 0 & 0.9753 & 0 & 0 & 0.001 & 0 & 0 & 0 & 0 & 0 \\ 0 & 0 & 0 & 0 & 0 & 0 & 0 & 0 & 0 & 0 & 0.9753 & 0 & 0 & 0.001 & 0 & 0 & 0 & 0 \\ 0 & 0 & 0 & 0 & 0 & 0 & 0 & 0 & 0 & 0 & 0 & 1 & 0 & 0 & 0.001 & 0 & 0 & 0 \\ 0 & 0.0001 & 0 & 0 & 0 & 0 & 0 & 0 & 0 & 0 & 0 & 0 & 1 & 0 & 0 & 0 & 0 & 0 \\ -0.0001 & 0 & 0 & 0 & 0 & 0 & 0 & 0 & 0 & 0 & 0 & 0 & 0 & 1 & 0 & 0 & 0 & 0 \\ 0 & 0 & 0 & 0 & 0 & 0 & 0 & 0 & 0 & 0 & 0 & 0 & 0 & 0 & 1 & 0 & 0 & 0 \\ 0 & 0.2422 & 0 & 0 & 0 & 0 & 0 & 0 & 0 & 0 & 0 & 0 & 0 & 0 & 0 & 0.9753 & 0 & 0 \\ -0.2422 & 0 & 0 & -0.0001 & 0 & 0 & 0 & 0 & 0 & 0 & 0 & 0 & 0 & 0 & 0 & 0 & 0.9753 & 0 \\ 0 & 0 & 0 & 0 & 0 & 0 & 0 & 0 & 0 & 0 & 0 & 0 & 0 & 0 & 0 & 0 & 0 & 0.9753 \end{bmatrix} \tag{4.19}$$

$$F(k) = \begin{bmatrix} 0 & 0 & 0 & 0 \\ 0 & 0 & 0 & 0 \\ 0 & 0 & 0 & 0 \\ 0 & 0.0012 & 0 & 0 \\ 0 & 0 & 0.00015 & 0 \\ 0 & 0 & 0 & 0 \\ 0 & 2.469 & 0 & 0 \\ 0 & 0 & 0.3010 & 0 \\ 0 & 0 & 0 & 1.668 \\ 0 & 0 & 0 & 0 \\ 0 & 0 & 0 & 0 \\ 0 & 0 & 0 & 0 \\ 0 & 0 & 0 & 0 \\ 0 & 0 & 0 & 0 \\ 0 & 0 & 0 & 0 \\ 0 & 0 & 0 & 0 \\ 0 & 0 & 0 & 0 \\ 0.8978 & 0 & 0 & 0 \end{bmatrix} ; \quad u(k) = \begin{bmatrix} F_T \\ \tau_\phi \\ \tau_\theta \\ \tau_\psi \end{bmatrix} \tag{4.20}$$

4.2 Discrete-Time Primary Control of Quadcopter

The primary controller of the Quadcopter is developed using the LQR technique in MATLAB for the discrete-time domain. In order to accomplish the objective, we need to design the feedback gain adequately, which will result in the system becoming stable. This proposed work is completely based on the discrete-time domain. So for that, the state space model has to be converted into the discrete-time domain, where all the state space variables are written in the form of samples instead of time, e.g., x(t) replaced with x(k).

To design the feedback gain using LQR approach, the Q and R weighted matrices are given in Eq. (4.21) depending on the required control effort. Q(k) will depend on the states of the system and R(k) will depend on the input of the system. Here dimension of the Q(k) matrix is 18×18 and the dimension of the R(k) matrix is 18×4.

$$Q = \begin{bmatrix} 1&0&0&0&0&0&0&0&0&0&0&0&0&0&0&0&0&0 \\ 0&1&0&0&0&0&0&0&0&0&0&0&0&0&0&0&0&0 \\ 0&0&1&0&0&0&0&0&0&0&0&0&0&0&0&0&0&0 \\ 0&0&0&1&0&0&0&0&0&0&0&0&0&0&0&0&0&0 \\ 0&0&0&0&1&0&0&0&0&0&0&0&0&0&0&0&0&0 \\ 0&0&0&0&0&1&0&0&0&0&0&0&0&0&0&0&0&0 \\ 0&0&0&0&0&0&1&0&0&0&0&0&0&0&0&0&0&0 \\ 0&0&0&0&0&0&0&1&0&0&0&0&0&0&0&0&0&0 \\ 0&0&0&0&0&0&0&0&1&0&0&0&0&0&0&0&0&0 \\ 0&0&0&0&0&0&0&0&0&1&0&0&0&0&0&0&0&0 \\ 0&0&0&0&0&0&0&0&0&0&1&0&0&0&0&0&0&0 \\ 0&0&0&0&0&0&0&0&0&0&0&1&0&0&0&0&0&0 \\ 0&0&0&0&0&0&0&0&0&0&0&0&1&0&0&0&0&0 \\ 0&0&0&0&0&0&0&0&0&0&0&0&0&1&0&0&0&0 \\ 0&0&0&0&0&0&0&0&0&0&0&0&0&0&1&0&0&0 \\ 0&0&0&0&0&0&0&0&0&0&0&0&0&0&0&1&0&0 \\ 0&0&0&0&0&0&0&0&0&0&0&0&0&0&0&0&1&0 \\ 0&0&0&0&0&0&0&0&0&0&0&0&0&0&0&0&0&1 \end{bmatrix}; \quad R = \begin{bmatrix} 1&0&0&0 \\ 0&1&0&0 \\ 0&0&1&0 \\ 0&0&0&1 \end{bmatrix} \tag{4.21}$$

The control law for the primary controller is given in Eq. (4.22),

$$u = -K_{dt} x(t) \tag{4.22}$$

where K_{dt} is the feedback gain having the dimension of 4×18, which can be calculated using Q and R weighted matrices.

4.2.1 Simulation Results of Discrete-Time Primary Control of Quadcopter

This section discusses the simulation results of primary control for a single Quadcopter using LQR control in the discrete-time domain. Using the LQR approach, we may get the simplified feedback gain K_{dt} as

$$K_{dt} = \begin{bmatrix} 0 & 0 & 0 & 0 & 0 & 0 & 0 & 0 & 0 & 0 & 0 & 1 & 0 & 0 & 15.1 & 0 & 0 & 1.5 \\ 21.4 & 0 & 0 & 8.9 & 0 & 0 & 0.94 & 0 & 0 & 0 & -1 & 0 & 0 & -2.4 & 0 & 0 & -0.24 & 0 \\ 0 & 39 & 0 & 0 & 28.1 & 0 & 0 & 2.5 & 0 & 1 & 0 & 0 & 3.1 & 0 & 0 & 0.3 & 0 & 0 \\ 0 & 0 & 1 & 0 & 0 & 2.21 & 0 & 0 & 0.48 & 0 & 0 & 0 & 0 & 0 & 0 & 0 & 0 & 0 \end{bmatrix} \tag{4.23}$$

Figure 4.1 shows the transitional motion, velocity, and acceleration in X, Y, and Z direction. Similarly, Fig. 4.2 shows rotational motion, angular velocity, and angular acceleration in the respective direction. The desired location of Quadcopter is given as $[X\ Y\ Z]^{\tau} = [15\ 25\ 35]^{\tau}$. Figure 4.3 shows how much control effort will be required to stabilize or maintain the desired output of the system. Figure 4.4 shows the Trajectory path followed by the Quadcopter as per the given input accurately.

4.3 Distributed Discrete-Time Sliding Mode (DDSM) Protocol for Formation of Quadcopter Swarm

In the discrete-time domain, a group of multiple Quadcopters is referred to as a discrete multi-agent system (DMAS). To establish the communication between those agents, the graph theory is used which is given in Sect. 3.2.4. With the help of the graph theory, the agents have their neighborhood agents information which is shared by the undirected graph topology. The DMAS of five Quadcopters is given below,

4.3.1 *Discrete Multi-agent System (DMAS)*

Considering the identical linear system for the ith agent as

$$x_i(k+1) = Ex_i(k) + Fu_i(k)$$

$$y_i(k) = Gx_i(k) + Hu_i(k) \tag{4.24}$$

According to the state space model given in Eqs. (4.19) and (4.20), $E \in \mathbb{R}^{18\times 18}$ and $F \in \mathbb{R}^{18\times 4}$ are the plant matrix and input matrix. The state vector $x_i(k) = [x_1(k)\ x_2(k)\ ....\ x_{18}(k)]^{\tau}$ and input vector $u_i(k) = [u_1(k)\ u_2(k)\ ....\ u_4(k)]^{\tau}$ [2].

Assumption 4.1 Considering the given system matrix pair(E, F) for ith agent is controllable and the DMAS under consideration does have a stationary, non-switching communication topology.

From Eq. (4.24), a global DMAS is derived as

$$X(k+1) = (I_N \otimes E)X(k) + (I_N \otimes F)u(k) \tag{4.25}$$

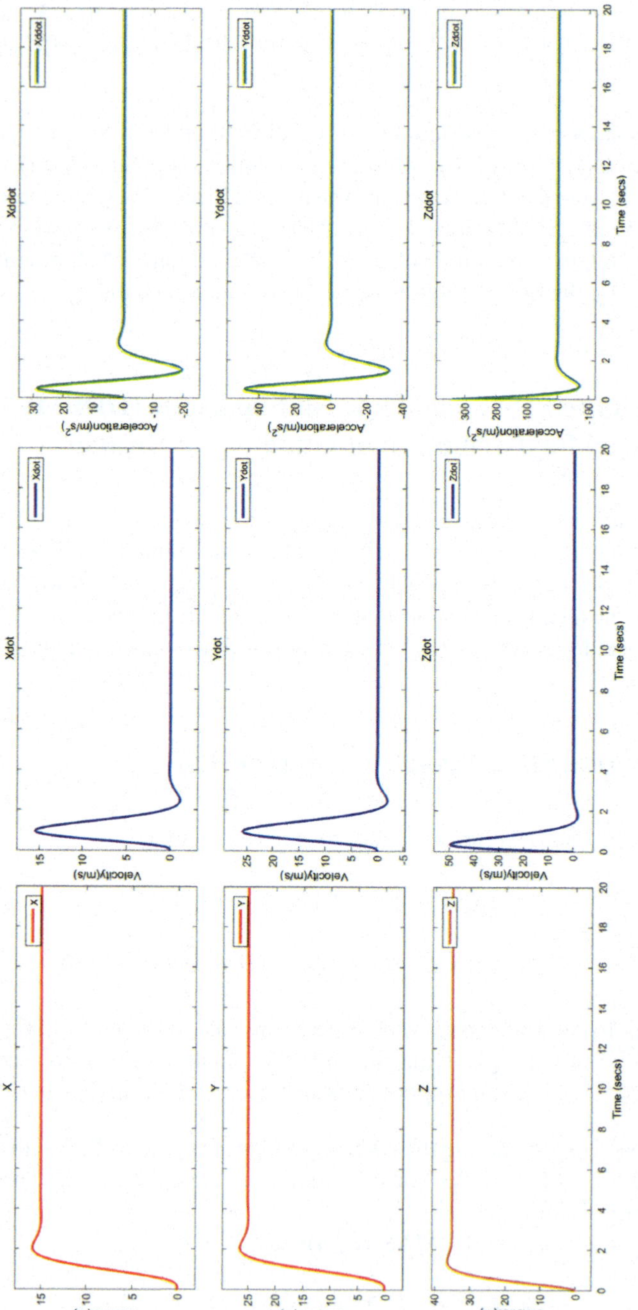

Fig. 4.1 Transitional motion of discrete-time primary control of Quadcopter

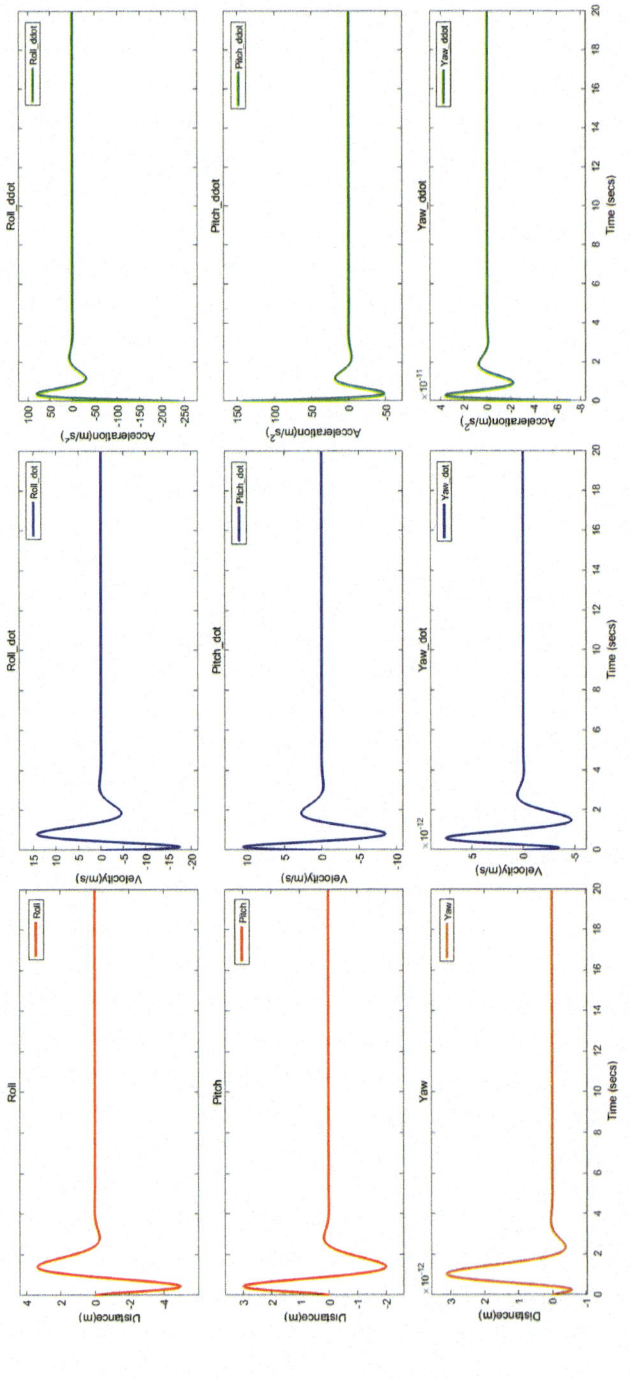

Fig. 4.2 Rotational motion of discrete-time primary control of Quadcopter

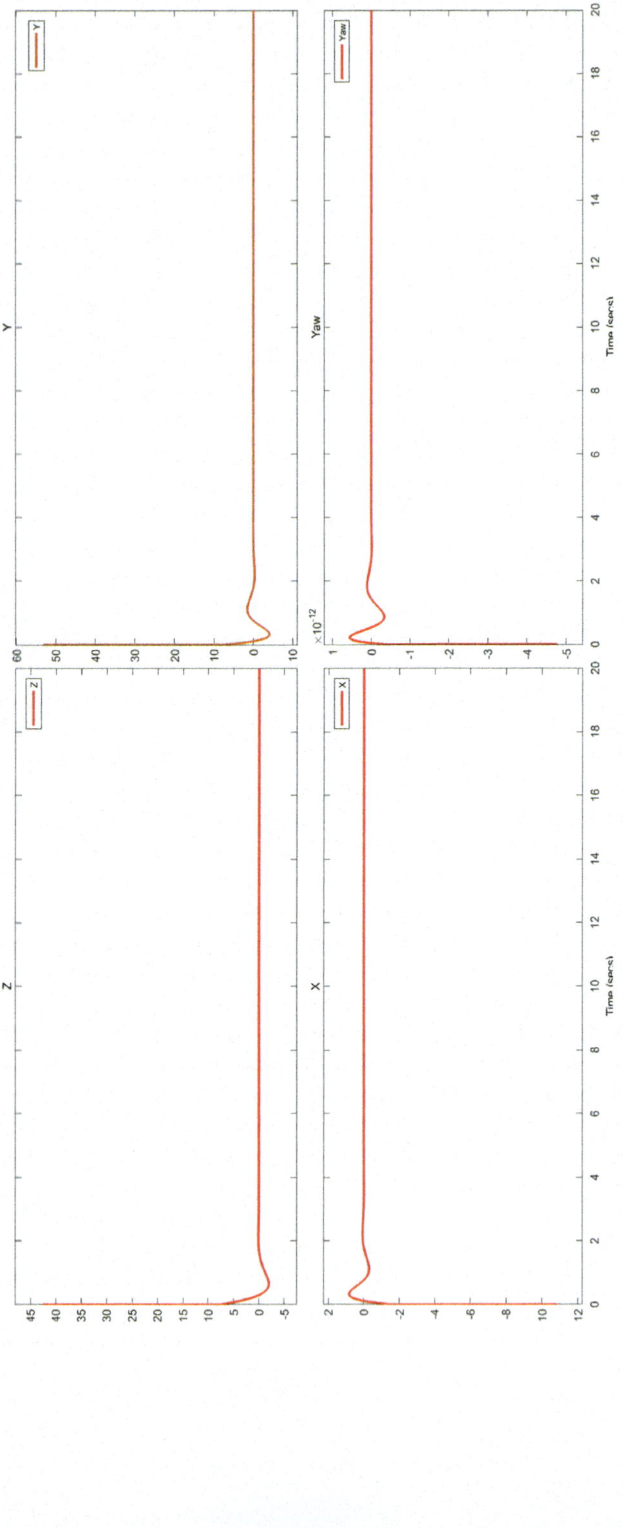

Fig. 4.3 Control effort of discrete-time primary control of Quadcopter

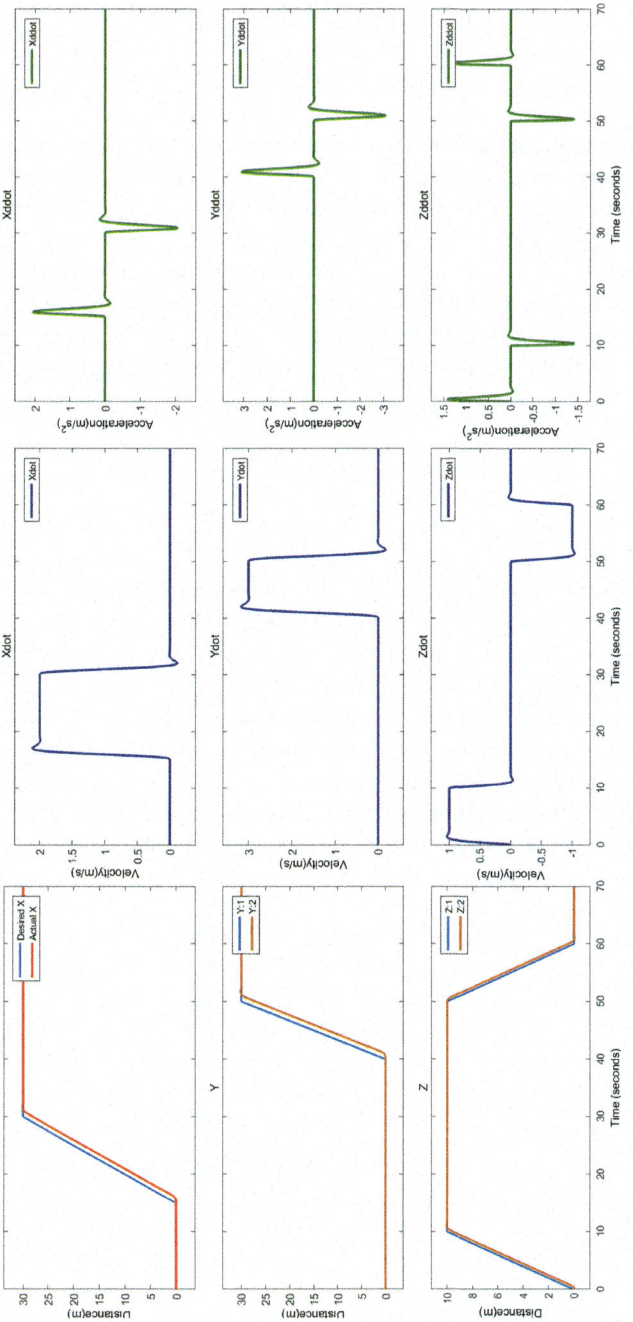

Fig. 4.4 Trajectory tracking of discrete-time primary control of Quadcopter

we may rewrite Eq. (4.24) as

$$X(k+1) = \bar{E}X(k) + \bar{F}u(k) \tag{4.26}$$

where $\otimes$ denotes the Kronecker product, N = 5, $\bar{E} = I_5 \otimes E$, and $\bar{F} = I_5 \otimes F$.

Formation Protocol for DMAS: For designing the distributed discrete-time sliding mode protocol using power rate reaching law, the formation pattern given in Fig. 4.7 is considered. For a Quadcopter system, there are four controlling parameters at the user end such as X, Y, Z, and Yaw position. Hence, the formation can be achieved using the linear parameters X, Y, and Z position. Consider a DMAS of 5 Quadcopter agents. Let the desired formation of the agents represented as $\mathbb{F} = (\mathbb{G}, \mathbb{Q})$ where $\mathbb{G}$ is the formation graph and $\mathbb{Q} = [q_1\ q_2 \ldots q_5]$ [3]. In the multiple Quadcopters system, the formation condition of the X position between agents i and j can be defined as

$$||X_i - X_j|| = d_{ij} \tag{4.27}$$

where d_{ij} is the desired distance between agents i and j. Similarly, the Y and Z positions can also be defined.

As per the formation pattern, the position of the leader agent is considered as the (0,0,0) for the (X, Y, Z) position. The immediate follower-1 (FA-1) and follower-2 (FA-2) have the distance d_{ij} with the leader agent in (X, Y, Z) position as (−42, −1, 0) and (2, −1, 0). Then the next follower-3 (FA-3) and follower-4 (FA-4) have the distance d_{ij} with leader agent in (X, Y, Z) position as (−1, −3, 0) and (1, −3, 0). The altitude of all the Quadcopters will remain the same or it can be presented as $d_{ij} = 0$ for all Quadcopters in the Z position.

While establishing the communication between all the agents we need to define The Laplacian $\mathbb{L}$, adjacency $\mathbb{A}$, degree $\mathbb{D}$, and diagonal matrices $\mathbb{B}$ are derived as

$$\mathbb{L} = \begin{bmatrix} 2 & 0 & 0 & 0 & 0 \\ -1 & 2 & 0 & 0 & 0 \\ -1 & 0 & 2 & 0 & 0 \\ 0 & -1 & 0 & -1 & 0 \\ 0 & 0 & -1 & 0 & 1 \end{bmatrix} \tag{4.28}$$

$$\mathbb{D} = \begin{bmatrix} 2 & 0 & 0 & 0 & 0 \\ 0 & 2 & 0 & 0 & 0 \\ 0 & 0 & 2 & 0 & 0 \\ 0 & 0 & 0 & 1 & 0 \\ 0 & 0 & 0 & 0 & 1 \end{bmatrix} \tag{4.29}$$

$$\mathbb{P} = \begin{bmatrix} 1\,0\,0\,0\,0 \\ 0\,0\,0\,0\,0 \\ 0\,0\,0\,0\,0 \\ 0\,0\,0\,0\,0 \\ 0\,0\,0\,0\,0 \end{bmatrix} \tag{4.30}$$

$$\mathbb{A} = \begin{bmatrix} 0\,0\,0\,0\,0 \\ 1\,0\,0\,0\,0 \\ 1\,0\,0\,0\,0 \\ 0\,1\,0\,0\,0 \\ 0\,0\,1\,0\,1 \end{bmatrix} \tag{4.31}$$

4.3.2 DDSM Formation Protocol for Quadcopter Swarm Using Power Rate Reaching Law

Since the power rate reaching law has been used for reducing the chattering effect and faster convergence as compared to Gao's reaching law. The distributed sliding mode using the power rate reaching law for the formation of a Quadcopter swarm is derived in Theorem 4.1.

Theorem 4.1 *The global homogeneous DMAS given in Eq. (4.26) is said to achieve the leader-follower formation in fixed time with the DDSM protocol using power rate reaching law given as*

$$\begin{aligned} u_p(k) = -(\bar{\Upsilon}_d\Gamma\bar{F})^{-1}(\bar{\Upsilon}_d\Gamma\bar{E}X(k) + \bar{\Upsilon}_d\Gamma(-1_5 \otimes \tilde{E}x_{r(k)} + \bar{d}_{ij}) \\ -(1-\lambda\tau)\bar{\mathbb{S}}(k) + \mu\tau|\bar{\mathbb{S}}(k)|^{\eta}sgn(\bar{\mathbb{S}}(k))) \end{aligned} \tag{4.32}$$

Proof Let us define the local formation error between the leader and agents for the DMAS as

$$\delta(k) = \sum[a_{ij}[X_i(k) - X_j(k)] + \mathbb{P}_i[X_i(k) - x_r(k)] + d_{ij}] \tag{4.33}$$

where $x_r(k)$ is the reference trajectory given to the leader agent and d_{ij} is the distance between the agents to achieve the formation.

From Eq. (4.33), the global error can be written as

$$\mathbb{E}(k) = (((I_5 + \bar{\mathbb{D}} + \bar{\mathbb{B}})^{-1})(\bar{\mathbb{L}} + \bar{\mathbb{B}}) \otimes I_{18})\hat{x} \tag{4.34}$$

Further expanding Eq. (4.34), we may get

$$\mathbb{E}(k) = (((I_5 + \bar{\mathbb{D}} + \bar{\mathbb{B}})^{-1})(\bar{\mathbb{L}} + \bar{\mathbb{B}}) \otimes I_{18})(X(k) - 1_5 \otimes x_{r(k)} + \bar{d}_{ij}) \tag{4.35}$$

where $\hat{x} = X(k) - 1_5 \otimes x_{r(k)} + \bar{d}_{ij}$ and $(I_5 + \bar{\mathbb{D}} + \bar{\mathbb{B}})^{-1}$ must be non-singular matrix. Equation (4.35) can be rewritten as

$$\mathbb{E}(k) = (\zeta \otimes I_{18})(X(k) - 1_5 \otimes x_{r(k)} + d_{ij}) \tag{4.36}$$

where $\zeta = (I_5 + \bar{\mathbb{D}} + \bar{\mathbb{B}})^{-1})(\bar{\mathbb{L}} + \bar{\mathbb{B}})$

$$\mathbb{E}(k) = \Gamma(X(k) - 1_5 \otimes x_{r(k)} + d_{ij}) \tag{4.37}$$

where $\Gamma = \zeta \otimes I_{18}$.

Now, define the sliding surface for the ith agent as

$$\mathbb{S}_i(k) = \Upsilon_d \delta(k) \tag{4.38}$$

where Υ_d is the sliding mode gain to be obtained using the Pole placement or LQR approach.

Then Eq. (4.38) can be further written in the global form as

$$\bar{\mathbb{S}}(k) = \bar{\Upsilon}_d \mathbb{E}(k) \tag{4.39}$$

where $\bar{\Upsilon}_d = I_5 \otimes \Upsilon_d$

Using Eq. (4.39), we may write

$$\bar{\mathbb{S}}(k) = \bar{\Upsilon}_d \Gamma(X(k) - 1_5 \otimes x_{r(k)} + \bar{d}_{ij}) \tag{4.40}$$

Further, advancing Eq. (4.40)

$$\bar{\mathbb{S}}(k+1) = \bar{\Upsilon}_d \Gamma(X(k+1) - 1_5 \otimes x_{r(k+1)} + \bar{d}_{ij}) \tag{4.41}$$

Substituting Eq. (4.24) into Eq. (4.41)

$$\bar{\mathbb{S}}(k+1) = \bar{\Upsilon}_d \Gamma(\bar{E}X(k) + \bar{F}u(k) - 1_5 \otimes x_{r(k)} + \bar{d}_{ij}) \tag{4.42}$$

Let us define the power rate reaching law for the ith agent as [4]

$$\mathbb{S}_i(k+1) = (1 - \lambda_i \tau_i)\mathbb{S}_i(k) - \mu_i \tau_i |\mathbb{S}_i(k)|^{\eta_i} sgn(\mathbb{S}_i(k)) \tag{4.43}$$

From Eq. (4.43), the power rate reaching law for the global homogeneous DMAS written as

$$\bar{\mathbb{S}}(k+1) = (1 - \lambda\tau)\bar{\mathbb{S}}(k) - \mu\tau|\bar{\mathbb{S}}(k)|^{\eta} sgn(\bar{\mathbb{S}}(k)) \tag{4.44}$$

Now, comparing Eq. (4.42) with Eq. (4.44), we may get the DDSM formation protocol using power rate reaching law for the homogeneous DMAS as

$$u_p(k) = -(\bar{\Upsilon}_d \Gamma \bar{F})^{-1}(\bar{\Upsilon}_d \Gamma \bar{E} X(k) + \bar{\Upsilon}_d \Gamma(-1_5 \otimes \tilde{E} x_{r(k)} + \bar{d}_{ij}) \\ -(1-\lambda\tau)\bar{\mathbb{S}}(k) + \mu\tau|\bar{\mathbb{S}}(k)|^{\eta} sgn(\bar{\mathbb{S}}(k))) \quad (4.45)$$

This completes the proof.

Let us consider the power rate reaching law in Eq. (4.43) to derive the quasi-sliding mode band (QSMB) as

$$\mathbb{S}_i(k+1) = (1-\lambda_i\tau_i)\mathbb{S}_i(k) - \mu_i\tau_i|\mathbb{S}_i(k)|^{\eta_i} sgn(\mathbb{S}_i(k)) \quad (4.46)$$

As per the definition of quasi-sliding mode band, the sign of the $\mathbb{S}_i(k+1)$ is opposite to $\mathbb{S}_i(k)$. So Eq. (4.46) can be written as

$$-\mathbb{S}_i(k) = (1-\lambda_i\tau_i)\mathbb{S}_i(k) - \mu_i\tau_i|\mathbb{S}_i(k)|^{\eta_i} sgn(\mathbb{S}_i(k)) \quad (4.47)$$

Further, we may get

$$2\mathbb{S}_i(k) = \lambda_i\tau_i\mathbb{S}_i(k) + \mu_i\tau_i|\mathbb{S}_i(k)|^{\eta_i} sgn(\mathbb{S}_i(k)) \quad (4.48)$$

Whenever the consensus error reaches to the QSMB, then considering the $\mathbb{S}_i(k) = \Delta_i$ and $sgn\mathbb{S}_i(k) = 1$ in Eq. (4.48) as

$$\Delta_i = [\frac{\mu_i\tau_i}{2-\lambda_i\tau_i}]^{\frac{1}{1-\eta_i}} \quad (4.49)$$

Equation (4.49) shows the chattering band for the sliding surface. While choosing the appropriate selection of η_i we can minimize the chattering effect and increase the speed of convergence.

4.3.3 *The Analysis of Number of Steps Required to Achieve the Formation of Quadcopter Swarm with DDSM Protocol Using Power Rate Reaching Law*

The next theorem presents the derivation of the total number of steps required to achieve the global formation for DDSM protocol using the power rate reaching law.

Theorem 4.2 ([5]) *The number of steps required to achieve the global formation of a homogeneous DMAS with DDSM protocol using power rate reaching law for the* ith *agent is given as*

$$[\hat{k}] = \frac{[ln(|\mathbb{S}_i(0)|^{1-\eta_i} + \frac{\mu_i}{\lambda_i}) - ln(\frac{\mu_i}{\lambda_i} + \Delta_i)]}{(1-\eta_i)\lambda_i\tau_i} \quad (4.50)$$

where $[\hat{k}]$ denotes the bounded maximum integer below the real value $\hat{k}$ and $\mathbb{S}_i(0)$ is the initial value of sliding surface.

Proof Let us consider the power rate reaching law form Eq. (4.43) for the ith agent and assume the τ_i is very small enough so that we may use the continuous-time derivative function

$$\dot{\mathbb{S}}_i = \lim_{\Delta \to 0} \frac{\mathbb{S}_i(t + \Delta(t)) - \mathbb{S}_i(t)}{\Delta(t)} \tag{4.51}$$

Rewriting Eq. (4.51), we may get

$$\dot{\mathbb{S}}_i = \frac{(\mathbb{S}_i(k+1) - \mathbb{S}_i(k))}{\tau_i} = -\lambda_i \mathbb{S}_i - \mu_i \mathbb{S}_i^{\eta_i} \tag{4.52}$$

Evaluating the reaching time for the error trajectories,

$$\frac{d\mathbb{S}_i}{dt} + \lambda_i \mathbb{S}_i = -\mu_i \mathbb{S}_i^{\eta_i} \tag{4.53}$$

Multiplying both side with $\mathbb{S}_i^{-\eta_i}(1$ - $\eta_i)$, we may get

$$(1 - \eta_i)\mathbb{S}_i^{-\eta_i} \frac{d\mathbb{S}_i}{dt} + (1 - \eta_i)\lambda_i \mathbb{S}_i^{1-\eta_i} = -(1 - \eta_i)\mu_i \tag{4.54}$$

Considering $\mathbb{Z} = \mathbb{S}_i^{1-\eta_i}$, we may get

$$\frac{d\mathbb{Z}}{dt} + (1 - \eta_i)\lambda_i \mathbb{Z} = -(1 - \eta_i)\mu_i \tag{4.55}$$

Now compare with the first-order differential equation defined as

$$\frac{d\mathbb{Z}}{dt} + \bar{\mathbb{P}}(t)\mathbb{Z} = \bar{\mathbb{Q}}(t) \tag{4.56}$$

The general solution of Eq. (4.56) is

$$\mathbb{Z} = ce^{\int \bar{\mathbb{P}}(t)dt} + e^{-\int \bar{\mathbb{P}}(t)dt} \int \bar{\mathbb{Q}}(t) e^{-\int \bar{\mathbb{P}}(t)dt} dt \tag{4.57}$$

Solving Eq. (4.57) after substituting the $\bar{\mathbb{P}}(t)$ and $\bar{\mathbb{Q}}(t)$, we may get

$$\mathbb{Z} = ce^{-(1-\eta_i)\lambda_i t} - \frac{\mu_i}{\lambda_i} \tag{4.58}$$

Substituting $\mathbb{Z}$, we may write

$$\mathbb{S}_i^{1-\eta_i} = ce^{-(1-\eta_i)\lambda_i t} - \frac{\mu_i}{\kappa_{1i}} \tag{4.59}$$

Taking t = 0 for finding the solution of c from Eq. (4.59) as

$$c = |\mathbb{S}_i(0)|^{(1-\eta_i)} + \frac{\mu_i}{\lambda_i} \tag{4.60}$$

substituting Eq. (4.59) into Eq. (4.60), we may get

$$\mathbb{S}_i^{1-\eta_i} = [|\mathbb{S}_i(0)|^{(1-\eta_i)} + \frac{\mu_i}{\lambda_i}]e^{-(1-\eta_i)}\lambda_i - \frac{\mu_i}{\lambda_i} \tag{4.61}$$

From Eq. (4.61), the QSMB for ith agent is defined as

$$\Delta_i = [|\mathbb{S}_i(0)|^{(1-\eta_i)} + \frac{\mu_i}{\lambda_i}]e^{-(1-\eta_i)}\lambda_i - \frac{\mu_i}{\lambda_i} \tag{4.62}$$

Calculating the reaching time, using the logarithmic function at both sides of Eq. (4.62) and we may get

$$t = \frac{\left\{\ln\left[|\mathbb{S}_i(0)|^{(1-\eta_i)} + \frac{\mu_i}{\lambda_i}\right] - \ln\left[\Delta_i + \frac{\mu_i}{\lambda_i}\right]\right\}}{(1-\eta_i)\,\lambda_i} \tag{4.63}$$

From Eq. (4.63), the discrete-time total number of steps evaluated from any $\mathbb{S}_i = \Delta_i$ given as

$$\hat{k} = t/\tau_i = \frac{\left\{\ln\left[|\mathbb{S}_i(0)|^{(1-\eta_i)} + \frac{\mu_i}{\lambda_i}\right] - \ln\left[\Delta_i + \frac{\mu_i}{\lambda_i}\right]\right\}}{(1-\eta_i)\,\lambda_i\tau_i} \tag{4.64}$$

From Eq. (4.64), appropriate selection of λ_i, μ_i, and η_i will reduce the chattering and improve the convergence speed of the formation.

This completes the proof.

Theorem 4.3 derives the global stability of leader-following formation of Quadcopter swarm using DDSM protocol using power rate reaching law.

Theorem 4.3 ([5]) *The global stability of the leader-following formation of Quadcopter swarm using DDSM formation protocol using power rate reaching law given in Eq. (4.45) is said to be stable, if the closed-loop error dynamics in Eq. (3.37) of DMAS drives toward the global sliding surface and maintained on it for any gain* λ *&* $\mu > 0$*,* $1 < 1 - \bar{\lambda}\tau < 0$*, and* $\eta > 0$ *provided the following condition is true:*

$$0 \le \bar{\beta} \le \bar{\mathbb{S}}(k)^{\tau}\bar{\mathbb{S}}(k) \tag{4.65}$$

where

$$\bar{\beta} = [(1-\lambda\tau)\bar{\mathbb{S}}(k) + \mu\tau|\bar{\mathbb{S}}(k)|^{\eta}sgn(\bar{\mathbb{S}}(k))]^{\tau} * \\ [(1-\lambda\tau)\bar{\mathbb{S}}(k) + \mu\tau|\bar{\mathbb{S}}(k)|^{\eta}sgn(\bar{\mathbb{S}}(k))] \tag{4.66}$$

Proof The global system of Quadcopter is given below

$$X(k+1) = (I_5 \otimes E)X(k) + (I_5 \otimes F)(u_{sw}(k) + u_{eq}(k)) \tag{4.67}$$

Firstly, define the global sliding surface as shown in Eq. (4.40)

$$\bar{\mathbb{S}}(k) = \bar{\Upsilon}_d \Gamma(X(k) - 1_5 \otimes x_{r(k)} + \bar{d}_{ij}) \tag{4.68}$$

Further, it may be written as

$$\bar{\mathbb{S}}(k+1) = \bar{\Upsilon}_d \Gamma(X(k+1) - 1_5 \otimes x_{r(k+1)} + \bar{d}_{ij}) \tag{4.69}$$

Let us define the Lyapunov function as

$$\mathcal{V}(k) = \bar{\mathbb{S}}^\tau \bar{\mathbb{S}} \tag{4.70}$$

differentiating Eq. (4.70)

$$\Delta\mathcal{V}(k) = \mathcal{V}(k+1) - \mathcal{V}(k) \tag{4.71}$$

$$\Delta\mathcal{V}(k) = \bar{\mathbb{S}}(k+1)^\tau \bar{\mathbb{S}}(k+1) - \bar{\mathbb{S}}(k)^\tau \bar{\mathbb{S}}(k) \tag{4.72}$$

Substituting $\bar{\mathbb{S}}(k+1)$ in Eq. (4.72), we may get

$$\Delta\mathcal{V}(k) = [\bar{\Upsilon}_d \Gamma(X(k+1) - 1_5 \otimes x_{r(k+1)} + \bar{d}_{ij})]^\tau * \\ [\bar{\Upsilon}_d \Gamma(X(k+1) - 1_5 \otimes x_{r(k+1)} + \bar{d}_{ij})] - \bar{\mathbb{S}}(k)^\tau \bar{\mathbb{S}}(k) \tag{4.73}$$

Substituting $X(k+1)$ in Eq. (4.73), we may get

$$\Delta\mathcal{V}(k) = [\bar{\Upsilon}_d \Gamma(\bar{E}X(k) + \bar{F}u(k) - 1_5 \otimes x_{r(k+1)} + \bar{d}_{ij})]^\tau * \\ [\bar{\Upsilon}_d \Gamma(\bar{E}X(k) + \bar{F}u(k) - 1_5 \otimes x_{r(k+1)} + \bar{d}_{ij})] - \bar{\mathbb{S}}(k)^\tau \bar{\mathbb{S}}(k) \tag{4.74}$$

Substituting the $u_p(k)$ in Eq. (4.45) into Eq. (4.74), we may get

$$\Delta\mathcal{V}(k) = [(1-\lambda\tau)\bar{\mathbb{S}}(k) - \mu\tau|\bar{\mathbb{S}}(k)|^\eta sgn(\bar{\mathbb{S}}(k))]^\tau * \\ [(1-\lambda\tau)\bar{\mathbb{S}}(k) - \mu\tau|\bar{\mathbb{S}}(k)|^\eta sgn(\bar{\mathbb{S}}(k))] - \bar{\mathbb{S}}(k)^\tau \bar{\mathbb{S}}(k) \tag{4.75}$$

Now denoting,

$$\bar{\beta} = [(1 - \lambda\tau)\bar{\mathbb{S}}(k) + \mu\tau|\bar{\mathbb{S}}(k)|^{\eta}sgn(\bar{\mathbb{S}}(k))]^{\tau} * \\ [(1 - \lambda\tau)\bar{\mathbb{S}}(k) + \mu\tau|\bar{\mathbb{S}}(k)|^{\eta}sgn(\bar{\mathbb{S}}(k))] \quad (4.76)$$

$$\Delta\mathcal{V}(k) = \bar{\beta} - \bar{\mathbb{S}}(k)^{\tau}\bar{\mathbb{S}}(k) \quad (4.77)$$

For stability purpose, the following condition holds true:

$$0 \leq \bar{\beta} \leq \bar{\mathbb{S}}(k)^{\tau}\bar{\mathbb{S}}(k) \quad (4.78)$$

From Eq. (4.78), the global stability of the DDSM formation protocol using power rate reaching law can be obtained by tuning the $\bar{\beta}$ so that the $\Delta\mathcal{V}(k) < 0$.

This completes the proof.

4.3.4 Simulation Results of DDSM Formation Protocol Using Power Rate Reaching Law

In this section, the simulation results are discussed of five homogeneous Quadcopters with the formation topology as shown in Fig. 4.5 with a DDSM protocol using the power rate reaching law. The sliding mode gain for the designed for the proposed protocol is calculated using the LQR approach. The gain for each agent λ and μ is chosen as [1 1 1 1] and [2 4 3 1], respectively. The value of η_i is 0.5 for the ith agent. Figures 4.6, 4.7 and 4.8 show the X, Y, and Z-axis position trajectory of five

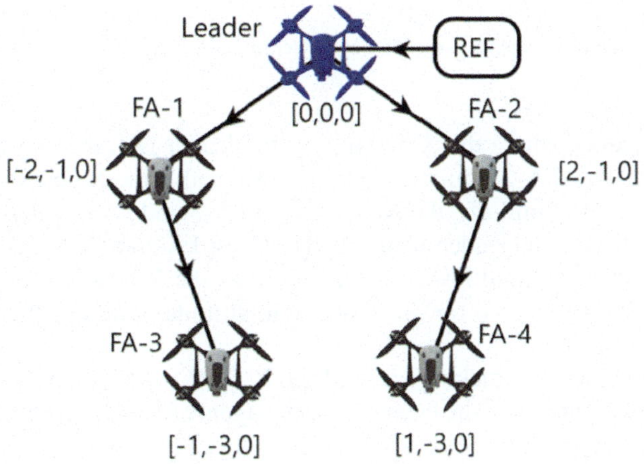

Fig. 4.5 Communication Topology

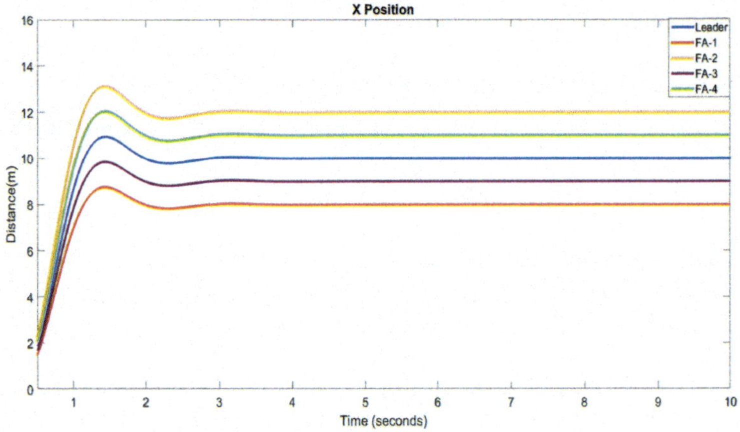

Fig. 4.6 X position of all agents using DDSM power rate reaching law

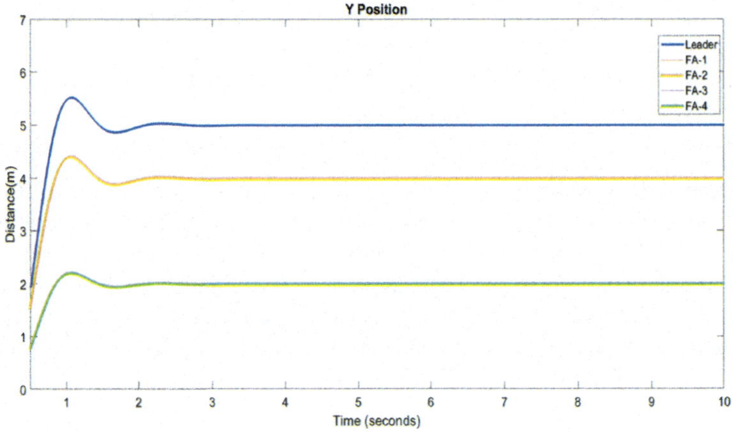

Fig. 4.7 Y position of all agents using DDSM power rate reaching law

Quadcopter agents, where the X position of the leader is 10m from the origin. As the d_{ij} for FA-1 and FA-2 from the leader is -2 m and 2, the X position of FA-1 and FA-2 is 8 and 12 m. Similarly, the X position of FA-3 and FA-4 is 9 and 11 m.

The Y position of the leader is 5m. As the d_{ij} for FA-1 and FA-2 is -1 m, so that the Y position of FA-1 and FA-2 is 4 m. Similarly, the Y position for the FA-2 and FA-3 is 2m. The d_{ij} is zero for the Z position, altitude of all agents is the same at 10 m.

Figure 4.9 shows the tracking error of the X, Y, Z, and Yaw position for all the agents. It can be noted that the error for all the agents reaches zero in a finite time.

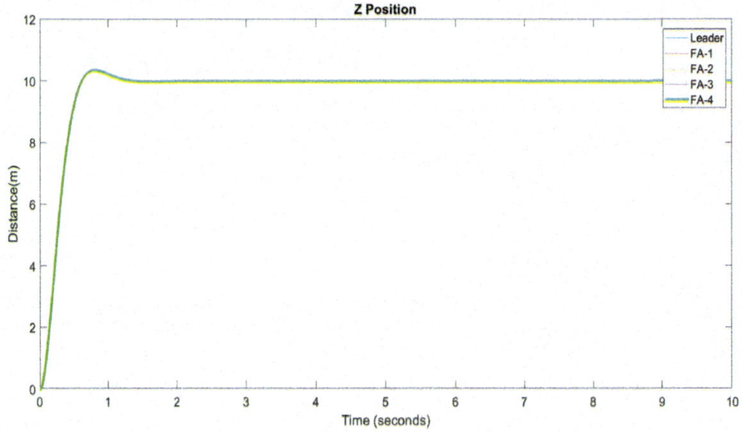

Fig. 4.8 Z position of all agents using DDSM power rate reaching law

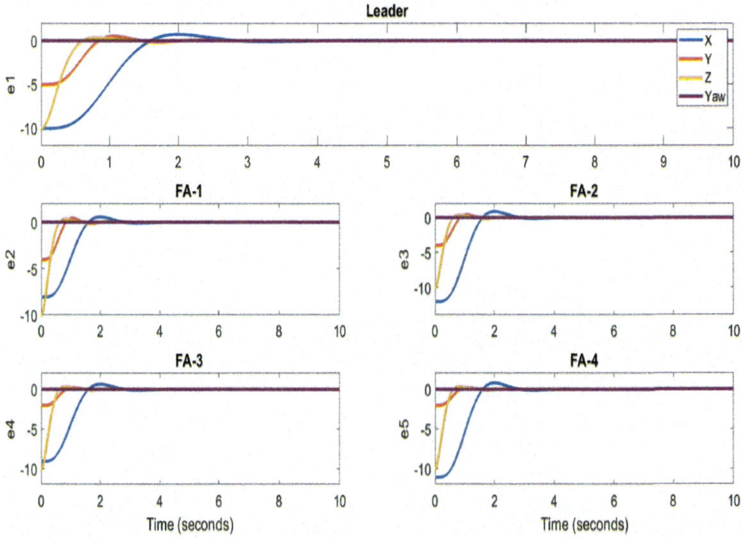

Fig. 4.9 Tracking error of all agents using DDSM power rate reaching law

Figure 4.10 shows the control efforts required to control the desired location generated by the individual agent. After settling at the desired location, the maximum control effort generated by the leader agent varies in the band of -0.00003 to $+0.00003$. Figure 4.11 shows the sliding surface of each Quadcopter. The sliding surface band is -2.2×10^{-6} to $+2.2 \times 10^{-6}$.

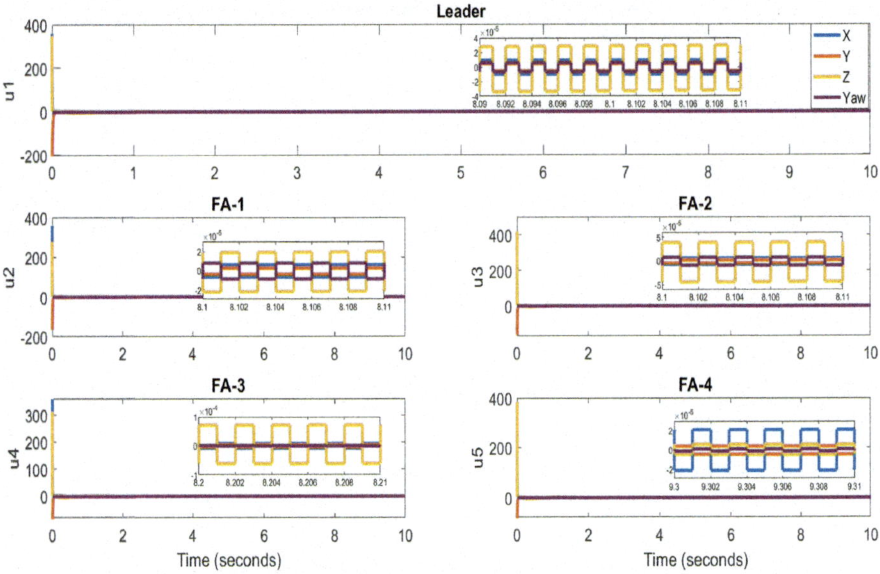

Fig. 4.10 Control efforts of all agents using DDSM power rate reaching law

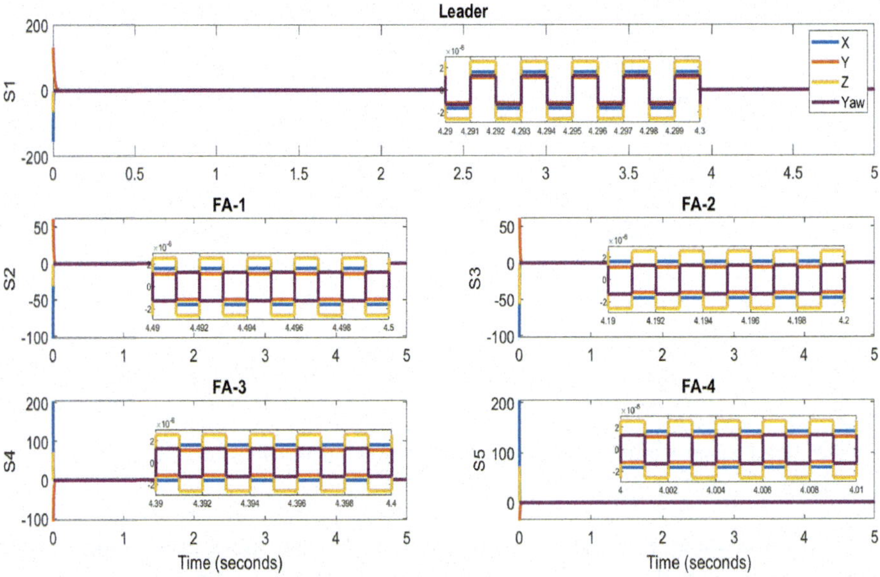

Fig. 4.11 Sliding surface of all agents using DDSM power rate reaching law

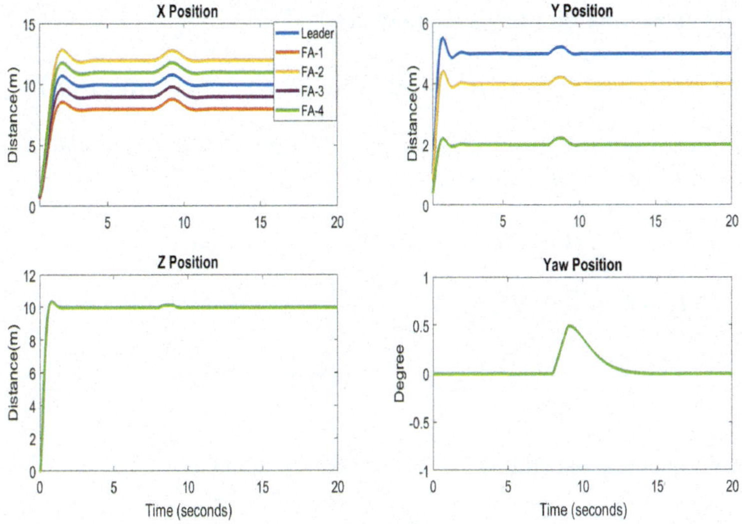

Fig. 4.12 Results of all agents under disturbance

4.4 Comparative Analysis of DDSM Using Power Rate Reaching Law with DDSM Using Gao's Reaching Law

To show the efficacy of the DDSM protocol using Power Rate reaching law, the results are compared with the classical DDSM protocol using Gao's reaching law.

Theorem 4.4 *The global homogeneous DMAS given in Eq. (4.26) is said to achieve the leader-follower formation in fixed time with the DDSM protocol using Gao's reaching law given as*

$$u_g(k) = -(\bar{\Upsilon}_d \Gamma \bar{F})^{-1}(\bar{\Upsilon}_d \Gamma \bar{E} X(k) + \bar{\Upsilon}_d \Gamma(-1_5 \otimes \tilde{E} x_{r(k)} + \bar{d}_{ij}) \\ -(1 - \kappa_1 \tau)\bar{\mathbb{S}}(k) + \kappa_2 \tau sgn(\bar{\mathbb{S}}(k))) \quad (4.79)$$

Proof Let us define the local formation error between the leader and agents for the DMAS as

$$\delta(k) = \sum [a_{ij}[X_i(k) - X_j(k)] + \mathbb{P}_i[X_i(k) - x_r(k)] + d_{ij}] \quad (4.80)$$

where $x_r(k)$ is the reference trajectory given to the leader agent. From Eq. (4.80), the global error can be written as

$$\mathbb{E}(k) = (((I_5 + \bar{\mathbb{D}} + \bar{\mathbb{B}})^{-1})(\bar{\mathbb{L}} + \bar{\mathbb{B}}) \otimes I_{18})\hat{x} \quad (4.81)$$

Further expanding Eq. (4.81), we may get

$$\mathbb{E}(k) = (((I_5 + \bar{\mathbb{D}} + \bar{\mathbb{B}})^{-1})(\bar{\mathbb{L}} + \bar{\mathbb{B}}) \otimes I_{18})(X(k) - 1_5 \otimes x_{r(k)} + \bar{d}_{ij}) \tag{4.82}$$

where $\hat{x} = X(k) - 1_5 \otimes x_{r(k)} + d_{ij}$ and $(I_5 + \bar{\mathbb{D}} + \bar{\mathbb{B}})^{-1}$ must be non-singular matrix. Equation (4.82) can be rewritten as

$$\mathbb{E}(k) = (\zeta \otimes I_{18})(X(k) - 1_5 \otimes x_{r(k)} + \bar{d}_{ij}) \tag{4.83}$$

where $\zeta = (I_5 + \bar{\mathbb{D}} + \bar{\mathbb{B}})^{-1})(\bar{\mathbb{L}} + \bar{\mathbb{B}})$

$$\mathbb{E}(k) = \Gamma(X(k) - 1_5 \otimes x_{r(k)} + \bar{d}_{ij}) \tag{4.84}$$

where $\Gamma = \zeta \otimes I_{18}$.

Now, define the local sliding surface for the ith agent as

$$\bar{\mathbb{S}}_i(k) = \Upsilon_d \delta(k) \tag{4.85}$$

where Υ_d is the sliding gain to be obtained using Pole placement or LQR approach.

Then Eq. (4.85) can be further written in the global form as

$$\bar{\mathbb{S}}(k) = \bar{\Upsilon}_d \mathbb{E}(k) \tag{4.86}$$

where $\bar{\Upsilon}_d = I_N \otimes \Upsilon_d$

Using Eq. (4.86), we may write

$$\bar{\mathbb{S}}(k) = \bar{\Upsilon}_d \Gamma(X(k) - 1_5 \otimes x_{r(k)} + \bar{d}_{ij}) \tag{4.87}$$

Further, advancing Eq. (4.88)

$$\bar{\mathbb{S}}(k+1) = \bar{\Upsilon}_d \Gamma(X(k+1) - 1_5 \otimes x_{r(k+1)} + \bar{d}_{ij}) \tag{4.88}$$

Substituting Eq. (4.26) into Eq. (4.89)

$$\bar{\mathbb{S}}(k+1) = \bar{\Upsilon}_d \Gamma(\bar{E}X(k) + \bar{F}u(k) - 1_5 \otimes x_{r(k)} + \bar{d}_{ij}) \tag{4.89}$$

Let us define Gao's reaching law for the ith agent as [6]

$$\bar{\mathbb{S}}_i(k+1) = (1 - \kappa_{1i}\tau_i)\bar{\mathbb{S}}_i(k) - \kappa_{2i}\tau_i sgn(\bar{\mathbb{S}}_i(k)) \tag{4.90}$$

From Eq. (4.90), Gao's reaching law for the global homogeneous DMAS written as

$$\bar{\mathbb{S}}(k+1) = (1 - \kappa_1\tau)\bar{\mathbb{S}}(k) - \kappa_2\tau sgn(\bar{\mathbb{S}}(k)) \tag{4.91}$$

where $\kappa_1 = I_5 \otimes \kappa_{1i}$ and $\kappa_2 = I_5 \otimes \kappa_{2i}$. Now, comparing Eq. (4.89) with Eq. (4.91), we may get the DDSM formation protocol using Gao's reaching law for the homogeneous DMAS as

$$u_g(k) = -(\bar{\Upsilon}_d \Gamma \bar{F})^{-1}(\bar{\Upsilon}_d \Gamma \bar{E} X(k) + \bar{\Upsilon}_d \Gamma(-1_5 \otimes \tilde{E} x_{r(k)} + \bar{d}_{ij}) \\ -(1 - \kappa_1 \tau)\bar{\mathbb{S}}(k) + \kappa_2 \tau sgn(\bar{\mathbb{S}}(k))) \quad (4.92)$$

This completes the proof.

Let us consider Gao's reaching law given in Eq. (4.86) to derive the quasi-sliding mode band (QSMB) as

$$\bar{\mathbb{S}}_i(k+1) = (1 - \kappa_{1i}\tau_i)\bar{\mathbb{S}}_i(k) - \kappa_{2i}\tau_i sgn(\bar{\mathbb{S}}_i(k)) \quad (4.93)$$

As per the definition of the quasi-sliding mode band, the sign of the $\bar{\mathbb{S}}_i(k+1)$ is opposite to $\bar{\mathbb{S}}_i(k)$. So Eq. (4.91) can be written as

$$-\bar{\mathbb{S}}_i(k) = (1 - \kappa_{1i}\tau_i)\bar{\mathbb{S}}_i(k) - \kappa_{2i}\tau_i sgn(\bar{\mathbb{S}}_i(k)) \quad (4.94)$$

Further, we may get

$$2\bar{\mathbb{S}}_i(k) = \kappa_{1i}\tau_i\bar{\mathbb{S}}_i(k) + \kappa_{2i}\tau_i sgn(\bar{\mathbb{S}}_i(k)) \quad (4.95)$$

Whenever the consensus error reaches to the QSMB, then considering the $\bar{\mathbb{S}}_i(k) = \Delta_i$ and $sgn(\bar{\mathbb{S}}_i(k)) = 1$ in Eq. (4.95) as

$$\Delta_i = [\frac{\kappa_{2i}\tau_i}{2 - \kappa_{1i}\tau_i}] \quad (4.96)$$

Equation (4.96) shows the chattering band for the sliding surface. While choosing the appropriate selection of κ_{1i} and κ_{2i}, we can minimize the chattering effect and increase the convergence speed.

4.4.1 The Analysis of Number of Steps Required to Achieve the Formation of Quadcopter Swarm with DDSM Protocol Using Gao's Reaching Law

Theorem 4.5 presents the total number of required steps to achieve the global formation of Quadcopter swarm with DDSM protocol using Gao's reaching law.

Theorem 4.5 ([7]) *The number of steps required for the global formation of a homogeneous DMAS with DDSM protocol using Gao's reaching law for the* ith *agent is given as*

$$[\hat{k}] = log_{(1-\kappa_{1i}\tau_i)}[\frac{\Delta_i \kappa_{1i} + \kappa_{2i}}{\kappa_{2i} + \kappa_{1i}|\bar{\mathbb{S}}_i(k)|}] \tag{4.97}$$

where $[\hat{k}]$ *denotes the bounded maximum integer below the real value* $\hat{k}$ *and* $\bar{\mathbb{S}}_i(0)$ *is the initial value of the sliding surface.*

Proof Based on the initial state of the sliding surface of the ith agent, there are two cases: $\bar{\mathbb{S}}_i(0) > 0$ and $\bar{\mathbb{S}}_i(0) < 0$.

Case 1: When $\bar{\mathbb{S}}_i(0) > 0$.

Consider Gao's reaching law from Eq. (4.91) as

$$\bar{\mathbb{S}}_i(k+1) = (1 - \kappa_{1i}\tau_i)\bar{\mathbb{S}}_i(k) - \kappa_{2i}\tau_i sgn(\bar{\mathbb{S}}_i(k)) \tag{4.98}$$

For k = 0, we may write

$$\bar{\mathbb{S}}_i(1) = (1 - \kappa_{1i}\tau_i)\bar{\mathbb{S}}_i(0) - \kappa_{2i}\tau_i sgn(\bar{\mathbb{S}}_i(0)) \tag{4.99}$$

and for k = 1,

$$\bar{\mathbb{S}}_i(2) = (1 - \kappa_{1i}\tau_i)\bar{\mathbb{S}}_i(1) - \kappa_{2i}\tau_i sgn(\bar{\mathbb{S}}_i(1)) \tag{4.100}$$

Substituting $\bar{\mathbb{S}}_i(1)$ into Eq. (4.100), we may get

$$\bar{\mathbb{S}}_i(2) = (1 - \kappa_{1i}\tau_i)[(1 - \kappa_{1i}\tau_i)\bar{\mathbb{S}}_i(0) - \kappa_{2i}\tau_i] - \kappa_{2i}\tau_i \tag{4.101}$$

Similarly for k = n, where n is the positive integer, we may get

$$\bar{\mathbb{S}}_i(n+1) = (1 - \kappa_{1i}\tau_i)^{n+1}\bar{\mathbb{S}}_i(0) - (1 - \kappa_{1i}\tau_i)^n\kappa_{2i}\tau_i - (1 - \kappa_{1i}\tau_i)\kappa_{2i}\tau_i - (1 - \kappa_{1i}\tau_i)^0\kappa_{2i}\tau_i \tag{4.102}$$

In the general form we can write

$$\bar{\mathbb{S}}_i(n+1) = (1 - \kappa_{1i}\tau_i)^{n+1}\bar{\mathbb{S}}_i(0) - \kappa_{2i}\tau_i \times \sum_{p=o}^{m}(1 - \kappa_{1i}\tau_i)^{n-p} \tag{4.103}$$

Equation (4.103) can be written as

$$\bar{\mathbb{S}}_i(n+1) = (1 - \kappa_{1i}\tau_i)^{n+1}\bar{\mathbb{S}}_i(0) - \kappa_{2i}\tau_i \times \frac{(1 - \kappa_{1i}\tau_i)^{n+1}}{\kappa_{1i}\tau_i} \tag{4.104}$$

Now, denoting n+1 = $\hat{k}$, we may write

$$\bar{\mathbb{S}}_i(\hat{k}) = (1 - \kappa_{1i}\tau_i)^{\hat{k}}\bar{\mathbb{S}}_i(0) - \kappa_{2i}\tau_i \times \frac{(1 - \kappa_{1i}\tau_i)^{\hat{k}}}{\kappa_{1i}\tau_i} \tag{4.105}$$

Now for derive the $\hat{k}$, consider the QSMB given in Eq. (4.96)

$$\Delta_i = (1 - \kappa_{1i}\tau_i)^{\hat{k}}\bar{\mathbb{S}}_i(0) - \kappa_{2i}\tau_i \times \frac{(1 - \kappa_{1i}\tau_i)^{\hat{k}}}{\kappa_{1i}\tau_i} \tag{4.106}$$

from Eq. (4.106),

$$(1 - \kappa_{1i}\tau_i)^{\hat{k}} = \frac{\Delta_i\kappa_{1i}\tau_i + \kappa_{2i}\tau_i}{\kappa_{2i}\tau_i + \kappa_{1i}\tau_i|\bar{\mathbb{S}}_i(0)|} \tag{4.107}$$

Applying log scale on both side of Eq. (4.107), we may get

$$[\hat{k}] = log_{(1-\kappa_{1i}\tau_i)}[\frac{\Delta_i\kappa_{1i} + \kappa_{2i}}{\kappa_{2i} + \kappa_{1i}|\bar{\mathbb{S}}_i(0)|}] \tag{4.108}$$

Case 2: $\bar{\mathbb{S}}_i(0) < 0$,

$$[\hat{k}] = log_{(1-\kappa_{1i}\tau_i)}[\frac{\Delta_i\kappa_{1i} + \kappa_{2i}}{\kappa_{2i} + \kappa_{1i}|\bar{\mathbb{S}}_i(0)|}] \tag{4.109}$$

Based on the initial state of the sliding surface variable pf the ith agent, the maximum $\hat{k} + 1$ steps required for a particular agent to synchronize with the leader are considered as a maximum number of steps required for the formation.

This completes the proof.

Theorem 4.6 derives the global stability of leader-following formation of Quadcopter swarm using DDSM protocol using Gao's reaching law.

Theorem 4.6 *The global stability of the leader-following formation of Quadcopter swarm using DDSM formation protocol using Gao's reaching law given in Eq. (4.90) is said to be stable, if the closed-loop error dynamics in Eq. (3.37) of DMAS drives toward the global sliding surface and maintained on it for any gain* κ_1 *&* $\kappa_2 > 0$, $1 < 1 - \bar{\kappa_1}\tau < 0$ *provided the following condition is true:*

$$0 \leq \bar{\beta} \leq \bar{\mathbb{S}}(k)^{\tau}\bar{\mathbb{S}}(k) \tag{4.110}$$

where

$$\bar{\beta} = [(1 - \kappa_1\tau)\bar{\mathbb{S}}(k) + \kappa_2\tau sgn(\bar{\mathbb{S}}(k))]^{\tau} * [(1 - \kappa_1\tau)\bar{\mathbb{S}}(k) + \kappa_2\tau\bar{\mathbb{S}}(k))] \tag{4.111}$$

Proof The global system of Quadcopter as given below

$$X(k+1) = (I_5 \otimes E)X(k) + (I_5 \otimes F)(u_{sw}(k) + u_{eq}(k)) \tag{4.112}$$

Firstly, define the global sliding surface as shown in Eq. (4.87)

$$\bar{\mathbb{S}}(k) = \Upsilon_d \Gamma (X(k) - 1_5 \otimes x_{r(k)} + \bar{d}_{ij}) \tag{4.113}$$

Further, it may be written as

$$\bar{\mathbb{S}}(k+1) = \Upsilon_d \Gamma (X(k+1) - 1_5 \otimes x_{r(k+1)} + \bar{d}_{ij}) \tag{4.114}$$

Let us define the Lyapunov function as

$$\mathcal{V}(k) = \bar{\mathbb{S}}^\tau \bar{\mathbb{S}} \tag{4.115}$$

differentiating Eq. (4.115)

$$\Delta \mathcal{V}(k) = \mathcal{V}(k+1) - \mathcal{V}(k) \tag{4.116}$$

$$\Delta \mathcal{V}(k) = \bar{\mathbb{S}}(k+1)^\tau \bar{\mathbb{S}}(k+1) - \bar{\mathbb{S}}(k)^\tau \bar{\mathbb{S}}(k) \tag{4.117}$$

Substituting $\bar{\mathbb{S}}(k+1)$ in Eq. (4.117), we may get

$$\begin{aligned} \Delta \mathcal{V}(k) = {} & [\Upsilon_d \Gamma (X(k+1) - 1_5 \otimes x_{r(k+1)} + \bar{d}_{ij})]^\tau * \\ & [\Upsilon_d \Gamma (X(k+1) - 1_5 \otimes x_{r(k+1)} + \bar{d}_{ij})] - \bar{\mathbb{S}}(k)^\tau \bar{\mathbb{S}}(k) \end{aligned} \tag{4.118}$$

Substituting $X(k+1)$ in Eq. (4.118), we may get

$$\begin{aligned} \Delta \mathcal{V}(k) = {} & [\Upsilon_d \Gamma (\bar{E} X(k) + \bar{F} u(k) - 1_5 \otimes x_{r(k+1)} + \bar{d}_{ij})]^\tau * \\ & [\Upsilon_d \Gamma (\bar{E} X(k) + \bar{F} u(k) - 1_5 \otimes x_{r(k+1)} + \bar{d}_{ij})] - \bar{\mathbb{S}}(k)^\tau \bar{\mathbb{S}}(k) \end{aligned} \tag{4.119}$$

Substituting the $u_g(k)$(k) in Eq. (4.92) into Eq. (4.119), we may get

$$\begin{aligned} \Delta \mathcal{V}(k) = {} & [(1 - \kappa_1 \tau) \bar{\mathbb{S}}(k) - \kappa_2 \tau sgn(\bar{\mathbb{S}}(k))]^\tau * \\ & [(1 - \kappa_1 \tau) \bar{\mathbb{S}}(k) - \kappa_2 \tau sgn(\bar{\mathbb{S}}(k))] - \bar{\mathbb{S}}(k)^\tau \bar{\mathbb{S}}(k) \end{aligned} \tag{4.120}$$

Now denoting,

$$\bar{\beta} = [(1 - \kappa_1 \tau) \bar{\mathbb{S}}(k) + \kappa_2 \tau sgn(\bar{\mathbb{S}}(k))]^\tau * [(1 - \kappa_1 \tau) \bar{\mathbb{S}}(k) + \kappa_2 \tau \bar{\mathbb{S}}(k))] \tag{4.121}$$

$$\Delta \mathcal{V}(k) = \bar{\beta} - \bar{\mathbb{S}}(k)^\tau \bar{\mathbb{S}}(k) \tag{4.122}$$

For stability purpose the following conditions hold true:

$$0 \leq \bar{\beta} \leq \bar{\mathbb{S}}(k)^\tau \bar{\mathbb{S}}(k) \tag{4.123}$$

From Eq. (4.123), the global stability of DDSM formation protocol using Gao's reaching law can be obtained by tuning the $\bar{\beta}$ so that the $\Delta\mathcal{V}(k) < 0$.

This completes the proof.

4.4.2 Simulation Results of DDSM Formation Protocol Using Gao's Reaching Law

Section 4.4.2 presents the comparison results of the formation of five Quadcopter swarms with DDSM protocol using Gao's reaching law. The position of the swarm in X, Y, and Z direction is shown in Figs. 4.13, 4.14, and 4.15. Similarly, the tracking error using Gao's reaching law is shown in Fig. 4.16. The required control efforts of all agents to achieve the formation of the swarm are given in Fig. 4.17. The control efforts of the swarm are varied in the band of -2×10^{-3} to $+2 \times 10^{-3}$. Figure 4.18 shows the sliding surface of all agents and the QSMB of the formation with DDSM protocol using Gao's reaching law is -2×10^{-4} to $+2 \times 10^{-4}$ given in the magnified window of the scale.

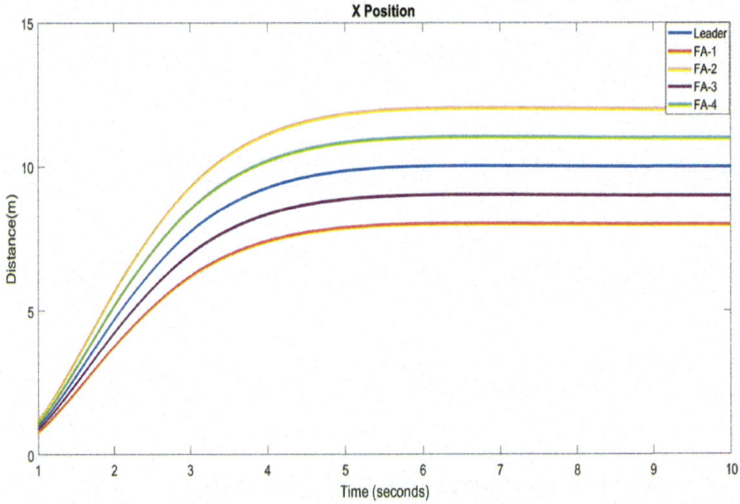

Fig. 4.13 X position of all agents using DDSM Gao's reaching law

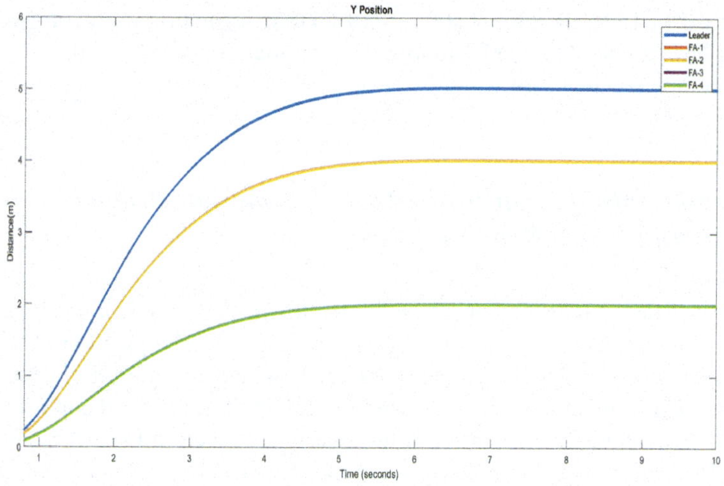

Fig. 4.14 Y position of all agents using DDSM Gao's reaching law

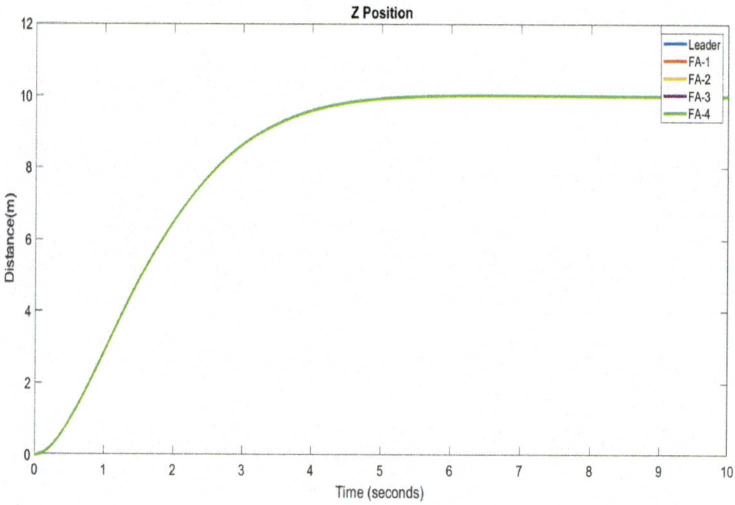

Fig. 4.15 Z position of all agents using DDSM Gao's reaching law

Also to evaluate the effectiveness of formation, the sliding surfaces and control efforts using power rate reaching law and Gao's reaching law are quantified as given in Tables 4.4 and 4.5. Moreover, the number of steps required for the global formation is also given in Table 4.6. From Table 4.6, it is observed that the formation achieved using DDSM protocol using the power rate reaching law is faster than Gao's reaching law.

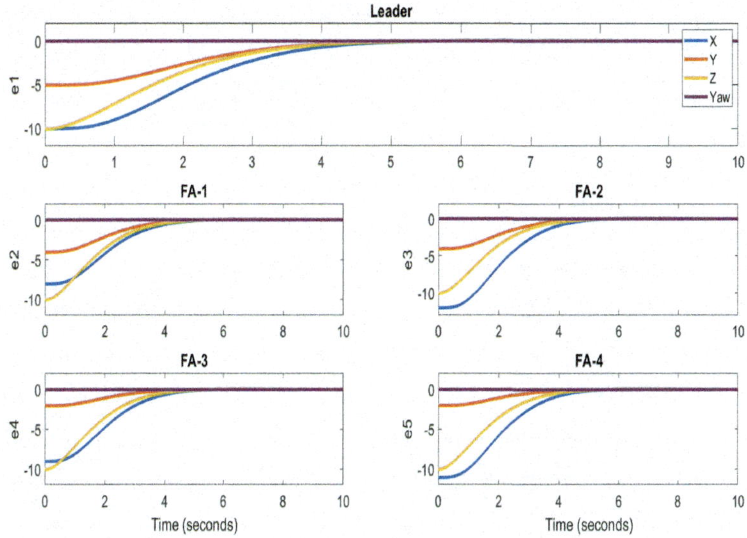

Fig. 4.16 Tracking error of all agents using DDSM Gao's reaching law

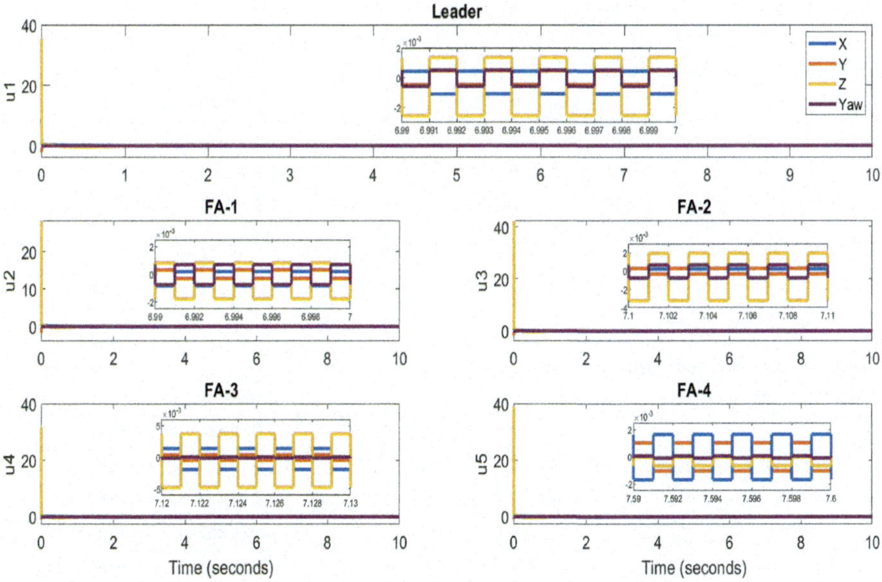

Fig. 4.17 Control efforts of all agents using DDSM Gao's reaching law

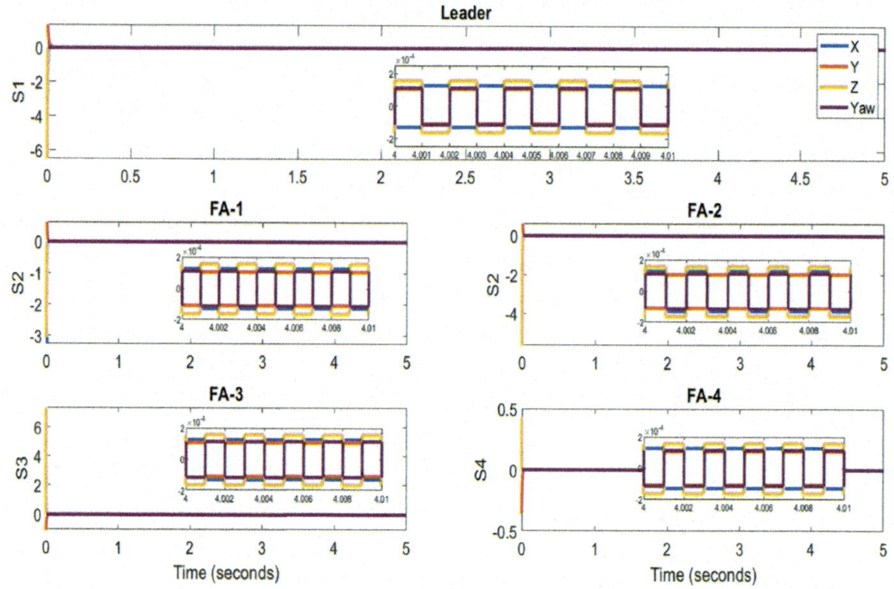

Fig. 4.18 Sliding surface of all agents using DDSM Gao's reaching law

Table 4.4 Comparison study of QSMB ($\triangle_i$)

QSMB	Gao's protocol	Power rate protocol
$\triangle_1$	3.2×10^{-4}	4.4×10^{-6}
$\triangle_2$	3.2×10^{-4}	4.4×10^{-6}
$\triangle_3$	3.2×10^{-4}	4.4×10^{-6}
$\triangle_4$	3.2×10^{-4}	4.4×10^{-6}
$\triangle_5$	3.2×10^{-4}	4.4×10^{-6}

Table 4.5 Comparison study of control effort

Control effort	Gao's protocol	Power rate protocol
u_1	−0.002 to +0.002	−0.00003 to +0.00003
u_2	−0.002 to +0.002	−0.00002 to +0.00002
u_3	−0.002 to +0.002	−0.00004 to +0.00004
u_4	−0.004 to +0.004	−0.00001 to +0.00001
u_5	−0.002 to +0.002	−0.000012 to +0.00002

Table 4.6 Number of steps required for the formation

$\hat{k}$	Gao's protocol	Power rate protocol
X Position	5060	3933
Y Position	3988	2787
Z Position	4757	2620

4.5 Conclusion

In this chapter, two distributed discrete-time sliding mode (DDSM) protocols using the power rate reaching law and Gao's reaching law are designed using the leader-following DMAS concept and the fixed graph theory to achieve the formation of Quadcopter swarm. The number of steps required for the formation is derived. The stability of the formation is also derived for both protocols using the Lyapunov function. It is inferred from the simulation results that the proposed formation protocol brings all Quadcopters in the given pattern in a finite time. Further, the performance of formation achieved using the power rate reaching law is compared with Gao's reaching law. The controller designed by the power rate reaching law gives a faster convergence than Gao's reaching law with a lesser chattering magnitude. The simulation results and the quantitative analysis show that the proposed formation control law outperforms the other controller.

References

1. https://www.quanser.com/products/autonomous-vehicles-research-studio/
2. Patel, K., Mehta, A.: Discrete-Time Sliding Mode Protocols for Discrete Multi-agent System, vol. 303, pp. 1–121. Springer (2021). https://doi.org/10.1007/978-981-15-6311-9
3. De Queiroz, M., Cai, X., Feemster, M.: Formation Control of Multi-agent Systems: A Graph Rigidity Approach. Wiley (2019). https://doi.org/10.1002/9781118887455
4. Devika, K.B., Thomas, S.: Power rate exponential reaching law for enhanced performance of sliding mode control. Int. J. Control, Autom. Syst. **15**, 2636–2645 (2017)
5. Patel, K., Mehta, A.: Discrete-time sliding mode protocols for leader-following consensus of discrete Multi-agent system with switching graph topology. European J. Control. ISSN 0947-3580. https://doi.org/10.1016/j.ejcon.2019.06.011
6. Gao, W., Wang, Y., Homaifa, A.: Discrete-time variable structure control systems. In: IEEE Transactions on Industrial Electronics, vol. 42, no. 2, pp. 117–122, Apr. 1995. https://doi.org/10.1109/41.370376
7. Patel, K., Mehta, A.: Fixed time steps discrete-time sliding mode consensus protocols for two degree of freedom helicopter systems. Trans. Inst. Meas. Control **43**(15), 3322–3334 (2021). https://doi.org/10.1177/01423312211004778

Chapter 5
Discrete-Time Second-Order Sliding Mode Protocol for Formation of Quadcopter Swarm

5.1 Introduction to Discrete Second-Order Sliding Model Protocol

In Chap. 5, the second-order sliding mode (SOSM) protocol which is known as the discrete-time super-twisting (DST) protocol has been designed. Over the conventional first-order discrete-time sliding mode control (FODSMC) method presented in Chap. 4, the discrete super-twisting protocol has plenty of advantages, particularly in situations where robust control with minimal chattering and adaptability to discrete-time systems are essential. Here are some key advantages of DST over FODSMC [1]:

- Chattering Suppression: In sliding mode control, chattering is a typical problem that is especially addressed by DST protocol. Chattering can cause control signals to oscillate at high frequencies, which is undesirable for mechanical systems. DST protocol effectively reduces chattering, resulting in smoother control actions.
- Reduced Control Effort: DST protocol often requires less control effort compared to FODSMC for achieving similar control performance. This can be advantageous in practical applications, as it reduces the energy consumption.
- Adaptability to Complex Systems: Complex systems with various states and inputs can be handled using the DST protocol. Systems with non-linear dynamics, time-varying parameters, and ambiguous disturbances can all benefit from using it.

A. Mehta and A. Modi, *Robust Sliding Mode Protocols for Formation of Quadcopter Swarm*, Studies in Systems, Decision and Control 521,
https://doi.org/10.1007/978-981-99-9726-8_5

5.2 Distributed Discrete-Time Second-Order Sliding Mode (DDSOSM) Super-Twisting (ST) Protocol for Formation of Quadcopter Swarm

Here, the DST protocol is used for the distributed controller in the formation whereas the LQR method is the same for the primary controller given in Sect. 4.2. To design the DDSOSM ST formation protocol, the pattern of the five Quadcopters as shown in Fig. 5.1 is considered. As per the formation pattern, the position of the leader agent is considered as the (0,0,0) for the (X, Y, Z) position. The immediate follower-1 (FA-1) and follower-2 (FA-2) have the distance d_{ij} with the leader agent in (X, Y, Z) position as (1,1,0) and (1,–1,0). Then the next follower-3 (FA-3) and follower-4 (FA-4) have the distance d_{ij} with leader agent in (X, Y, Z) position as (2,2,0) and (2,–2,0).

The altitude of all the Quadcopters will remain the same or it can be presented as $d_{ij} = 0$ for all Quadcopters in the Z position.

Theorem 5.1 presents the global formation for five homogeneous Quadcopters using the DDSOSM ST protocol.

Theorem 5.1 ([2]) *The global homogeneous DMAS is said to achieve the leader-follower formation in fixed time with the discrete-time super-twisting protocol given as*

$$u_s(k) = -[(\bar{\Upsilon}_d \Gamma \bar{F})^{-1}(\bar{\Upsilon}_d \Gamma \bar{E} X(k) + \bar{\Upsilon}_d \Gamma(-1_5 \otimes x_{r(k+1)} + \bar{d}_{ij}))] - \bar{\kappa}_1 \sqrt{|\bar{\mathbb{S}}(k)|} sgn(\bar{\mathbb{S}}(k)) + \omega(k) \quad (5.1)$$

Proof Let us define the local consensus error between the leader and agents for the DMAS as

$$\delta(k) = \sum [a_{ij}[X_i(k) - X_j(k)] + \mathbb{P}_n[X_i(k) - X_r(k)] + d_{ij}] \quad (5.2)$$

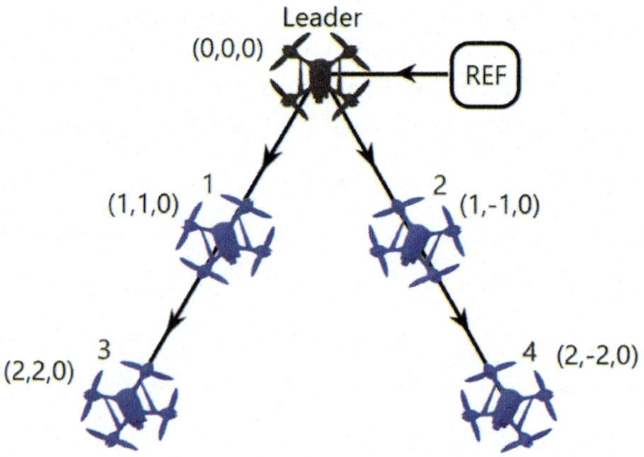

Fig. 5.1 Communication Topology

where $X_r(k)$ is the reference trajectory given to the leader agent. From Eq. (5.2), the global error can be written as

$$\mathbb{E}(k) = (((I_5 + \bar{\mathbb{D}} + \bar{\mathbb{B}})^{-1})(\bar{\mathbb{L}} + \bar{\mathbb{B}}) \otimes I_{18})\hat{x} \tag{5.3}$$

Further expanding Eq. (5.3), we may get

$$\mathbb{E}(k) = (((I_5 + \bar{\mathbb{D}} + \bar{\mathbb{B}})^{-1})(\bar{\mathbb{L}} + \bar{\mathbb{B}}) \otimes I_{18})(X(k) - 1_5 \otimes x_{r(k)} + \bar{d}_{ij}) \tag{5.4}$$

where $\otimes$ is Kronecker product and $\hat{x} = X(k) - 1_5 \otimes x_{r(k)} + \bar{d}_{ij}$ and $(I_5 + \bar{\mathbb{D}} + \bar{\mathbb{B}})^{-1}$ must be non-singular matrix.

Equation (5.4) can be rewritten as

$$\mathbb{E}(k) = \Gamma\hat{x} \tag{5.5}$$

Substituting $\hat{x}$,

$$\mathbb{E}(k) = \Gamma(X(k) - 1_5 \otimes x_{r(k)} + \bar{d}_{ij}) \tag{5.6}$$

where $\Gamma = \zeta \otimes I_{18}$.

Now, define the local sliding surface for the ith agent as

$$\mathbb{S}_i(k) = \Upsilon_d \delta(k) \tag{5.7}$$

where Υ_d is the sliding gain to be obtained using the Pole placement or LQR approach.

Then Eq. (5.7) can be written in the form global system as

$$\bar{\mathbb{S}}(k) = \bar{\Upsilon}_d \mathbb{E}(k) \tag{5.8}$$

where $\bar{\Upsilon}_d = I_5 \otimes \Upsilon_d$

Using Eq. (5.8), we may write

$$\bar{\mathbb{S}}(k) = \bar{\Upsilon}_d \Gamma(X(k) - 1_5 \otimes x_{r(k)} + \bar{d}_{ij}) \tag{5.9}$$

Further, advancing Eq. (5.9)

$$\bar{\mathbb{S}}(k+1) = \bar{\Upsilon}_d \Gamma(X(k+1) - 1_5 \otimes x_{r(k+1)} + \bar{d}_{ij}) \tag{5.10}$$

Substituting X(k+1) into Eq. (5.10)

$$\bar{\mathbb{S}}(k+1) = \bar{\Upsilon}_d \Gamma(\bar{E}X(k) + \bar{F}u(k) - 1_5 \otimes x_{r(k+1)} + \bar{d}_{ij}) \tag{5.11}$$

The discrete super-twisting control protocol has two parts, $u_s(k)$ switching law, and $u_{eq}(k)$ equivalent law. Taking $\mathbb{S}(k+1) = 0$ and resulting $u_{eq}(k)$ is defined as

$$0 = \bar{\Upsilon}_d \Gamma(\bar{E}X(k) + \bar{F}u_{eq}(k) - 1_5 \otimes x_{r(k+1)} + \bar{d}_{ij}) \tag{5.12}$$

After solving Eq. (5.12), we may get

$$u_{eq}(k) = [-(\bar{\Upsilon}_d \Gamma \bar{F})^{-1}[\bar{\Upsilon}_d \Gamma \bar{E} X(k) + \bar{\Upsilon}_d \Gamma (-1_5 \otimes x_{r(k+1)} + \bar{d}_{ij})]] \quad (5.13)$$

Let us define the discrete super-twisting protocol-based DSMC as [3]

$$u_{isw}(k) = -\kappa_1 \sqrt{|\mathbb{S}_i(k)|} sgn(\mathbb{S}_i(k)) + \omega(k)$$

$$\omega(k+1) = \omega(k) - \kappa_2 \tau sgn(\mathbb{S}_i(k)) \quad (5.14)$$

where κ_1 , $\kappa_2 > 0$.

The discrete super-twisting protocol can be written in global form as

$$u_{sw}(k) = -\bar{\kappa}_1 \sqrt{|\bar{\mathbb{S}}(k)|} sgn(\bar{\mathbb{S}}(k)) + \omega(k)$$

$$\omega(k+1) = \omega(k) - \bar{\kappa}_2 \tau sgn(\bar{\mathbb{S}}(k)) \quad (5.15)$$

Hence, we have the control law as

$$u_s(k) = u_{sw}(k) + u_{eq}(k) \quad (5.16)$$

Now, using Eqs. (5.13) and (5.15), we may get the DDSOSM ST formation protocol for the homogeneous DMAS of Quadcopter swarm as

$$u_s(k) = -[(\bar{\Upsilon}_d \Gamma \bar{F})^{-1}(\bar{\Upsilon}_d \Gamma \bar{E} X(k) + \bar{\Upsilon}_d \Gamma (-1_5 \otimes x_{r(k+1)} + \bar{d}_{ij}))] - \bar{\kappa}_1 \sqrt{|\bar{\mathbb{S}}(k)|} sgn(\bar{\mathbb{S}}(k)) + \omega(k) \quad (5.17)$$

This completes the proof.

The next theorem presents the global stability of DMAS of a homogeneous Quadcopter swarm using the DDSOSM ST protocol.

Theorem 5.2 ([4]) *The global stability of the leader-following formation of Quadcopter swarm using DDSOSM ST formation protocol given in Eq. (5.17) is said to be stable, if the closed-loop error dynamics in Eq. (5.6) of DMAS drives toward the global sliding surface and maintained on it for any gain* $\bar{\kappa}_1$ & $\bar{\kappa}_2 > 0$ *provided the following condition is true:*

$$0 \leq \bar{\beta} \leq \bar{\mathbb{S}}(k)^{\tau} \bar{\mathbb{S}}(k) \quad (5.18)$$

where

$$\bar{\beta} = [-\bar{\kappa}_1 \sqrt{|\bar{\mathbb{S}}(k)|} sgn(\bar{\mathbb{S}}(k)) + \omega(k)]^{\tau} * [-\bar{\kappa}_1 \sqrt{|\bar{\mathbb{S}}(k)|} sgn(\bar{\mathbb{S}}(k)) + \omega(k)] \quad (5.19)$$

Proof The global system of Quadcopters is given below

$$X(k+1) = (I_5 \otimes E)X(k) + (I_5 \otimes F)(u_{sw}(k) + u_{eq}(k)) \tag{5.20}$$

Firstly, define the global sliding surface as shown in Eq. (5.8)

$$\bar{\mathbb{S}}(k) = \bar{\Upsilon}_d \Gamma(X(k) - 1_5 \otimes x_{r(k)}) \tag{5.21}$$

Further, it may be written as

$$\bar{\mathbb{S}}(k+1) = \bar{\Upsilon}_d \Gamma(X(k+1) - 1_5 \otimes x_{r(k+1)}) \tag{5.22}$$

Let us define the Lyapunov function as

$$\mathcal{V}(k) = \bar{\mathbb{S}}^\tau \bar{\mathbb{S}} \tag{5.23}$$

differentiating Eq. (5.23)

$$\Delta\mathcal{V}(k) = \mathcal{V}(k+1) - \mathcal{V}(k) \tag{5.24}$$

$$\Delta\mathcal{V}(k) = \bar{\mathbb{S}}(k+1)^\tau \bar{\mathbb{S}}(k+1) - \bar{\mathbb{S}}(k)^\tau \bar{\mathbb{S}}(k) \tag{5.25}$$

Substituting $\bar{\mathbb{S}}(k+1)$ in Eq. (5.25), we may get

$$\Delta\mathcal{V}(k) = [\bar{\Upsilon}_d \Gamma(X(k+1) - 1_5 \otimes x_{r(k+1)} + \bar{d}_{ij})]^\tau * \\ [\bar{\Upsilon}_d \Gamma(X(k+1) - 1_5 \otimes x_{r(k+1)} + \bar{d}_{ij})] - \bar{\mathbb{S}}(k)^\tau \bar{\mathbb{S}}(k) \tag{5.26}$$

Substituting $X(k+1)$ in Eq. (5.26), we may get

$$\Delta\mathcal{V}(k) = [\bar{\Upsilon}_d \Gamma(\bar{E}X(k) + \bar{F}u(k) - 1_5 \otimes x_{r(k+1)} + \bar{d}_{ij})]^\tau * \\ [\bar{\Upsilon}_d \Gamma(\bar{E}X(k) + \bar{F}u(k) - 1_5 \otimes x_{r(k+1)} + \bar{d}_{ij})] - \bar{\mathbb{S}}(k)^\tau \bar{\mathbb{S}}(k) \tag{5.27}$$

Substituting the u_s(k) in into Eq. (5.27), we may get

$$\Delta\mathcal{V}(k) = [-\bar{\kappa}_1 \sqrt{|\bar{\mathbb{S}}(k)|} sgn(\bar{\mathbb{S}}(k)) + \omega(k)]^\tau * \\ [-\bar{\kappa}_1 \sqrt{|\bar{\mathbb{S}}(k)|} sgn(\bar{\mathbb{S}}(k)) + \omega(k)] - \bar{\mathbb{S}}(k)^\tau \bar{\mathbb{S}}(k) \tag{5.28}$$

Now denoting,

$$\bar{\beta} = [-\bar{\kappa}_1 \sqrt{|\bar{\mathbb{S}}(k)|} sgn(\bar{\mathbb{S}}(k)) + \omega(k)]^\tau * \\ [-\bar{\kappa}_1 \sqrt{|\bar{\mathbb{S}}(k)|} sgn(\bar{\mathbb{S}}(k)) + \omega(k))] \tag{5.29}$$

$$\Delta\mathscr{V}(k) = \bar{\beta} - \bar{\mathbb{S}}(k)^{\tau}\bar{\mathbb{S}}(k) \tag{5.30}$$

For stability purpose, the following condition holds true:

$$0 \leq \bar{\beta} \leq \bar{\mathbb{S}}(k)^{\tau}\bar{\mathbb{S}}(k) \tag{5.31}$$

From the above condition given in Eq. (5.31), it can be understood that the value of $\bar{\beta}$ must be bounded by the $\bar{\mathbb{S}}(k)^{\tau}\bar{\mathbb{S}}(k)$. The $\bar{\beta}$ is dependent on the sliding mode gain $\bar{\Upsilon}_d$, and the super-twisting constants κ_1 and κ_2. So, the Quadcopter swarm will remain stable during the whole operation with the proper selection of the sliding mode gain and the positive values of constants that satisfy the above condition.

5.2.1 Simulation Results of DDSOSM ST Formation Protocol

The simulation analysis is carried out for a swarm of five Quadcopters using the DDSOSM ST formation protocol. All agents shared their information to the neighborhood agent as per the formation pattern shown in Fig. 5.1. The required parameter to evaluate the global error such as the Laplacian $\mathbb{L}$, adjacency $\mathbb{A}$, degree $\mathbb{D}$, and pinning $\mathbb{P}$ matrices are derived as

$$\mathbb{L} = \begin{bmatrix} 2 & 0 & 0 & 0 & 0 \\ -1 & 2 & 0 & 0 & 0 \\ -1 & 0 & 2 & 0 & 0 \\ 0 & -1 & 0 & -1 & 0 \\ 0 & 0 & -1 & 0 & 1 \end{bmatrix} \tag{5.32}$$

$$\mathbb{D} = \begin{bmatrix} 2 & 0 & 0 & 0 & 0 \\ 0 & 2 & 0 & 0 & 0 \\ 0 & 0 & 2 & 0 & 0 \\ 0 & 0 & 0 & 1 & 0 \\ 0 & 0 & 0 & 0 & 1 \end{bmatrix} \tag{5.33}$$

$$\mathbb{P} = \begin{bmatrix} 1 & 0 & 0 & 0 & 0 \\ 0 & 0 & 0 & 0 & 0 \\ 0 & 0 & 0 & 0 & 0 \\ 0 & 0 & 0 & 0 & 0 \\ 0 & 0 & 0 & 0 & 0 \end{bmatrix} \tag{5.34}$$

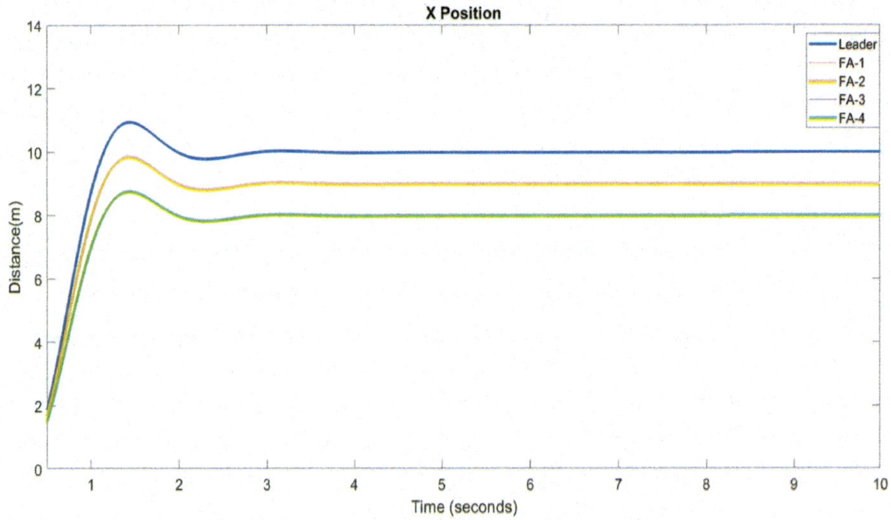

Fig. 5.2 Tracking of X position of all agents using DDSOSM ST protocol

$$\mathbb{A} = \begin{bmatrix} 0 & 0 & 0 & 0 & 0 \\ 1 & 0 & 0 & 0 & 0 \\ 1 & 0 & 0 & 0 & 0 \\ 0 & 1 & 0 & 0 & 0 \\ 0 & 0 & 1 & 0 & 1 \end{bmatrix} \tag{5.35}$$

To design the sliding surface, the sliding mode gain $\bar{\mathbb{S}}(k)$ is calculated using the LQR approach. The constant gains $\bar{\kappa}_1$ and $\bar{\kappa}_2$ chosen as 0.11 and 0.015, respectively. The output in the direction of X-, Y-, and Z-axis for all agents are shown in Figs. 5.2, 5.3, and 5.4, respectively. Figure 5.2 shows the X-axis position trajectory of five Quadcopter agents, where the X position of the leader is 10 m from the origin. As the d_{ij} is 1 m for FA-1 and FA-2 from the leader, the X position of FA-1 and FA-2 is 9 m. Similarly, the X position of FA-3 and FA-4 is 8 m. Figure 5.3 shows the Y position of five Quadcopters. The d_{ij} for between all agents 1 m in Y position. As the d_{ij} is zero for the Z position, the altitude of all agents is the same at 10 m shown in Fig. 5.4.

Figure 5.5 shows the tracking error of the X, Y, Z, and Yaw positions for all the agents. It can be noted that the error for all the agents reaches zero in a finite time. Figure 5.6 shows the control efforts required to achieving and maintaining the formation of Five Quadcopters. The chattering in control effort is shown as a magnified window. After settling at the desired location, efforts generated by the leader agent vary in the band of –0.001 to +0.001. Figure 5.7 shows the sliding surface of each Quadcopter. The sliding surface band is –0.0001 to +0.0001.

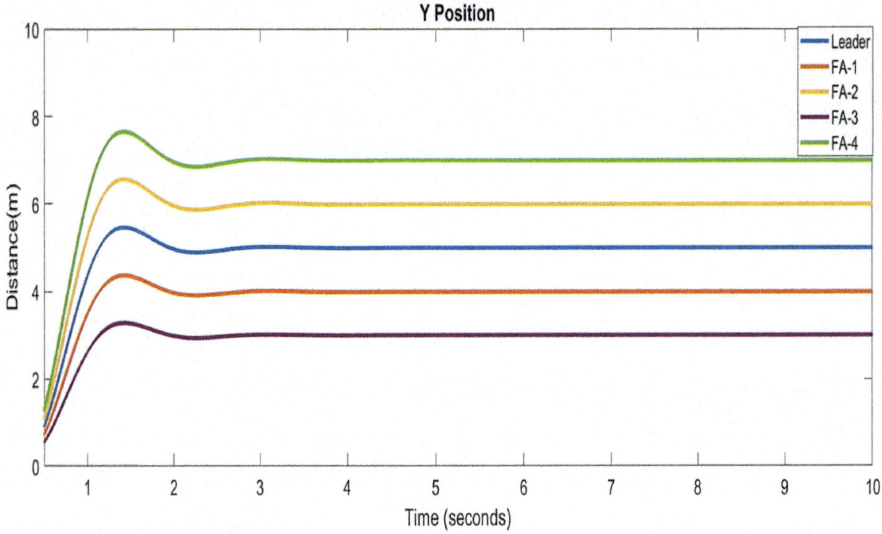

Fig. 5.3 Tracking of Y position of all agents using DDSOSM ST protocol

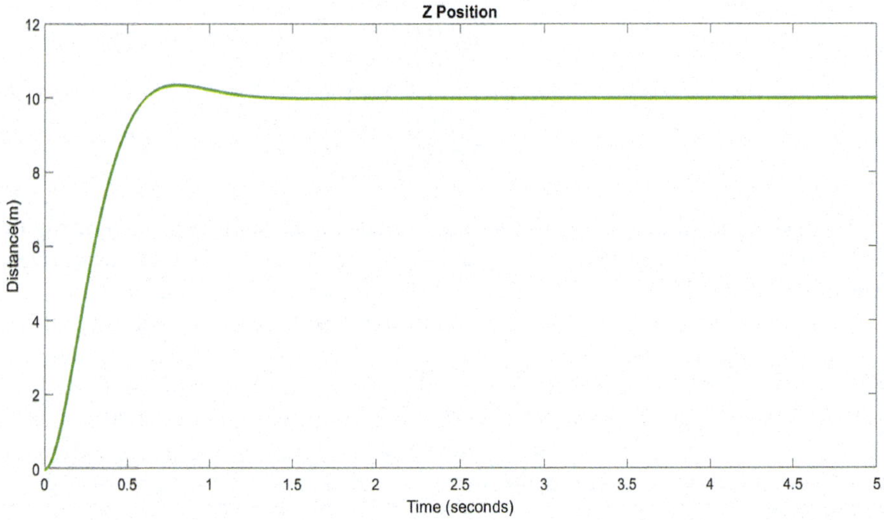

Fig. 5.4 Tracking of Z position of all agents using DDSOSM ST protocol

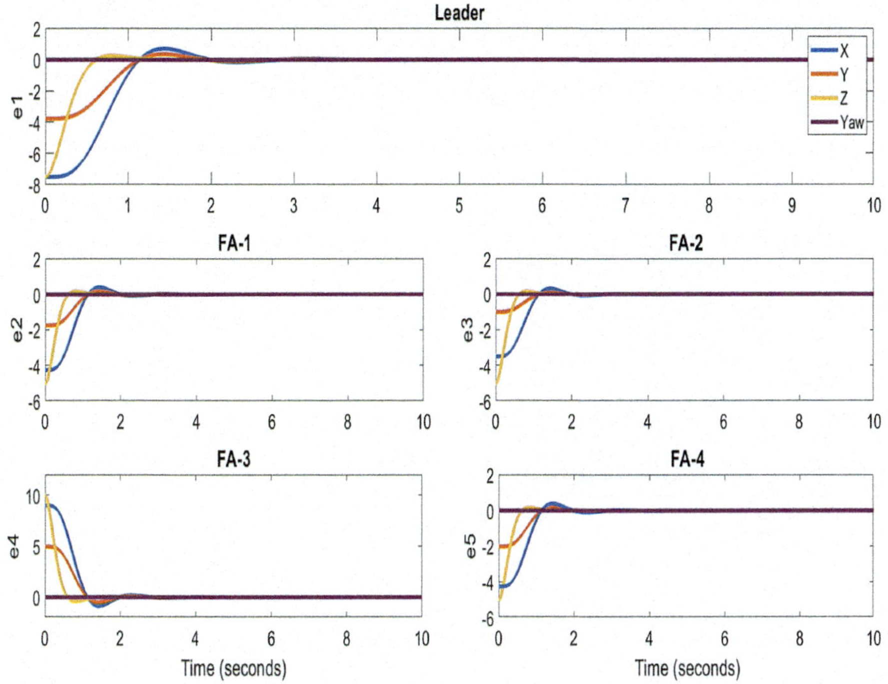

Fig. 5.5 Tracking error of all agents using DDSOSM ST protocol

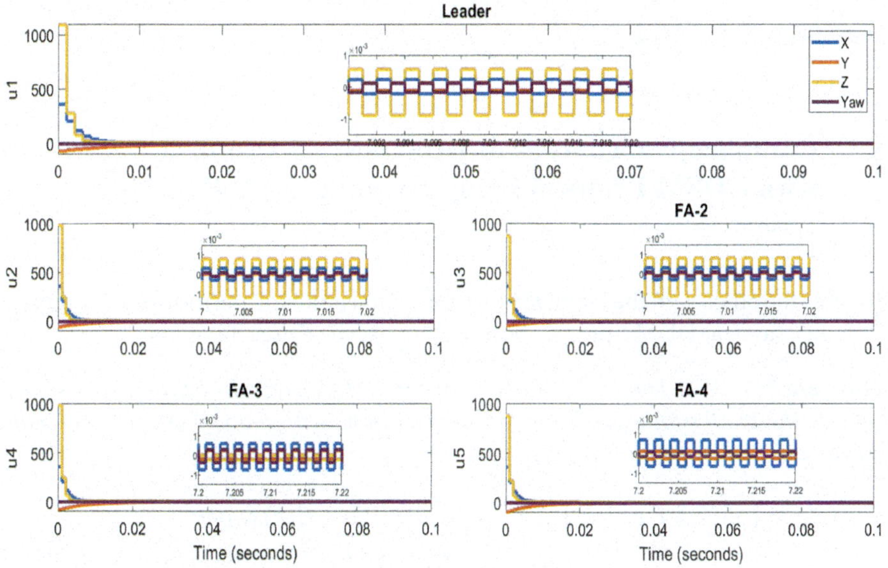

Fig. 5.6 Control effort of all agents using DDSOSM ST protocol

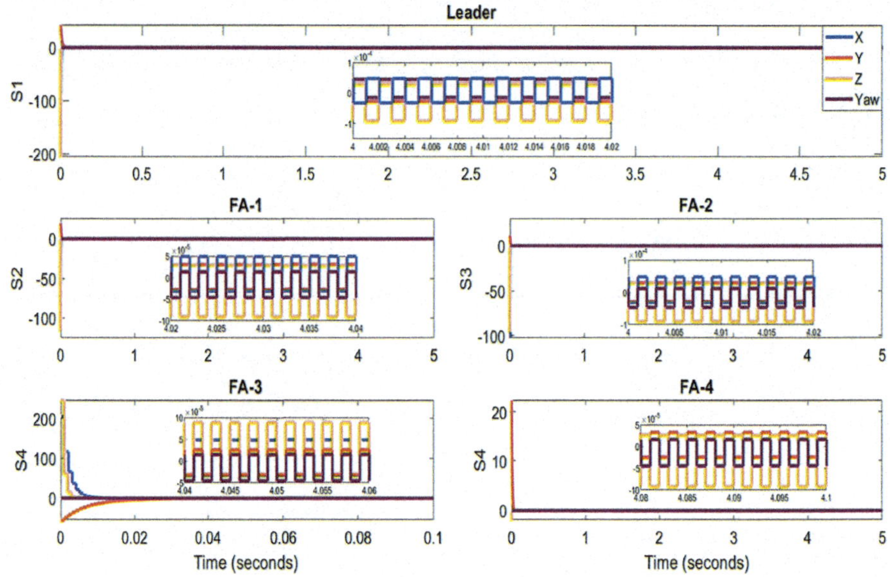

Fig. 5.7 Sliding surface of all agents using DDSOSM ST protocol

To show the efficacy of the DDSOSM ST formation protocol performance, the simulation results are compared with the two other classical discrete-time sliding mode protocols, namely, Gao's reaching law and Bartoszewicz reaching law given in the Sects. 5.3 and 5.4.

5.3 Comparative Analysis of DDSOSM ST Protocol with DDSM Protocol Using Bartoszewicz's Reaching Law

To show the efficacy of the DDSOSM ST formation protocol, the results are compared with the classical DDSM protocol using Bartoszewicz's reaching law.

Theorem 5.3 ([5]) *The global homogeneous DMAS given in Eq. (4.26) is said to achieve the leader-follower formation in fixed time with the DDSM protocol using Bartoszewicz's reaching law given as*

$$u_b(K) = [-(\bar{\Upsilon}_d \Gamma \bar{F})^{-1}[\bar{\Upsilon}_d \Gamma \bar{E} X(k) + \bar{\Upsilon}_d \Gamma (-1_5 \otimes E(k) x_{r(k)}) - I_{18} \otimes (d(k) - d(0)) - \bar{\mathbb{S}}_d(k+1)]] \quad (5.36)$$

Proof Let us define the local formation error between the leader and agents for the DMAS as

$$\delta(k) = \sum [a_{ij}[X_i(k) - X_j(k)] + \mathbb{P}_i[X_i(k) - X_r(k)] + d_{ij}] \tag{5.37}$$

where $X_r(k)$ is the reference trajectory given to the leader agent. From Eq. (5.37), the global error can be written as

$$\mathbb{E}(k) = (((I_5 + \bar{\mathbb{D}} + \bar{\mathbb{B}})^{-1})(\bar{\mathbb{L}} + \bar{\mathbb{B}}) \otimes I_{18})\hat{x} \tag{5.38}$$

Further expanding Eq. (5.38), we may get

$$\mathbb{E}(k) = (((I_5 + \bar{\mathbb{D}} + \bar{\mathbb{B}})^{-1})(\bar{\mathbb{L}} + \bar{\mathbb{B}}) \otimes I_{18})(X(k) - 1_5 \otimes x_{r(k)} + \bar{d}_{ij}) \tag{5.39}$$

where $\hat{x} = X(k) - 1_5 \otimes x_{r(k)} + \bar{d}_{ij}$ and $(I_5 + \bar{\mathbb{D}} + \bar{\mathbb{B}})^{-1}$ must be non-singular matrix.

Equation (5.39) can be rewritten as

$$\mathbb{E}(k) = \Gamma(X(k) - 1_5 \otimes x_{r(k)} + \bar{d}_{ij}) \tag{5.40}$$

where $\zeta = (I_5 + \bar{\mathbb{D}} + \bar{\mathbb{B}})^{-1})(\bar{\mathbb{L}} + \bar{\mathbb{B}})$

$$\mathbb{E}(k) = \Gamma(X(k) - 1_5 \otimes x_{r(k)} + \bar{d}_{ij}) \tag{5.41}$$

where $\Gamma = \zeta \otimes I_{18}$.

Now, define the local sliding surface for the ith agent as

$$\mathbb{S}_i(k) = \Upsilon_d \delta(k) \tag{5.42}$$

where Υ_d is the sliding gain to be obtained using Pole placement or LQR approach.

Then Eq. (5.42) can be further written in the global form as

$$\bar{\mathbb{S}}(k) = \bar{\Upsilon}_d \mathbb{E}(k) \tag{5.43}$$

where $\bar{\Upsilon}_d = I_5 \otimes \Upsilon_d$

Using Eq. (5.43), we may write

$$\bar{\mathbb{S}}(k) = \bar{\Upsilon}_d \Gamma(X(k) - 1_5 \otimes x_{r(k)} + \bar{d}_{ij}) \tag{5.44}$$

Further, advancing Eq. (5.44)

$$\bar{\mathbb{S}}(k+1) = \bar{\Upsilon}_d \Gamma(X(k+1) - 1_5 \otimes x_{r(k+1)} + \bar{d}_{ij}) \tag{5.45}$$

Substituting X(k+1) into Eq. (5.45)

$$\bar{\mathbb{S}}(k+1) = \bar{\Upsilon}_d \Gamma(\bar{E}X(k) + \bar{F}u(k) - 1_N \otimes x_{r(k)} + \bar{d}_{ij}) \tag{5.46}$$

Let us define Bartoszewicz's reaching law for the ith agent as

$$\bar{\mathbb{S}}_i(k+1) = d(k) - d(0) + \bar{\mathbb{S}}_{id}(k+1) \tag{5.47}$$

From Eq. (5.47), Bartoszewicz's reaching law for the global homogeneous DMAS written as

$$\bar{\mathbb{S}}(k+1) = I_{18} \otimes (d(k) - d(0)) + \bar{\mathbb{S}}_d(k+1) \tag{5.48}$$

Now, comparing Eq. (5.46) with Eq. (5.48), we may get the DDSM formation protocol for the homogeneous DMAS as

$$u_b(K) = [-(\bar{\Upsilon}_d \Gamma \bar{F})^{-1}[\bar{\Upsilon}_d \Gamma \bar{E} X(k) + \bar{\Upsilon}_d \Gamma(-1_5 \otimes E(k)x_{r(k)}) \\ -I_{18} \otimes (d(k) - d(0)) - \bar{\mathbb{S}}_d(k+1)]] \tag{5.49}$$

This completes the proof.

Theorem 5.4 presents the global stability of the formation of Quadcopter swarm using DDSM using Bartoszewicz's reaching law.

Theorem 5.4 *The global stability of the leader-following formation of Quadcopter swarm using DDSM formation protocol using Bartoszewicz's reaching law given in Eq. (5.49) is said to be stable, if the closed-loop error dynamics in Eq. (5.41) of DMAS drives toward the global sliding surface and maintained on it for any gain* $d_l < d_k < d_0$, $\mathbb{S}_d(k) = 0$ *for* $k \geq k^*$ *provided the following condition is true:*

$$0 \leq \bar{\beta} \leq \bar{\mathbb{S}}(k)^{\tau} \bar{\mathbb{S}}(k) \tag{5.50}$$

where

$$\bar{\beta} = [I_{18} \otimes (d(k) - d(0)) + \bar{\mathbb{S}}_d(k+1)]^{\tau} * [I_{18} \otimes (d(k) - d(0)) + \bar{\mathbb{S}}_d(k+1)] \tag{5.51}$$

Proof The global system of Quadcopter as given below

$$X(k+1) = (I_5 \otimes E)X(k) + (I_5 \otimes F)(u_{sw}(k) + u_{eq}(k)) \tag{5.52}$$

Firstly, define the global sliding surface as shown in Eq. (5.44)

$$\bar{\mathbb{S}}(k) = \bar{\Upsilon}_d \Gamma(X(k) - 1_5 \otimes x_{r(k)} + \bar{d}_{ij}) \tag{5.53}$$

Further, it may be written as

$$\bar{\mathbb{S}}(k+1) = \bar{\Upsilon}_d \Gamma(X(k+1) - 1_5 \otimes x_{r(k+1)} + \bar{d}_{ij}) \tag{5.54}$$

Let us define the Lyapunov function as

$$\mathcal{V}(k) = \bar{\mathbb{S}}^{\tau}\bar{\mathbb{S}} \tag{5.55}$$

differentiating Eq. (5.55)

$$\Delta\mathcal{V}(k) = \mathcal{V}(k+1) - \mathcal{V}(k) \tag{5.56}$$

$$\Delta\mathcal{V}(k) = \bar{\mathbb{S}}(k+1)^{\tau}\bar{\mathbb{S}}(k+1) - \bar{\mathbb{S}}(k)^{\tau}\bar{\mathbb{S}}(k) \tag{5.57}$$

Substituting $\bar{\mathbb{S}}(k+1)$ in Eq. (5.57), we may get

$$\begin{aligned}\Delta\mathcal{V}(k) = {} & [\bar{\Upsilon}_d\Gamma(X(k+1) - 1_5 \otimes x_{r(k+1)} + \bar{d}_{ij})]^{\tau} * \\ & [\bar{\Upsilon}_d\Gamma(X(k+1) - 1_5 \otimes x_{r(k+1)} + \bar{d}_{ij})] - \bar{\mathbb{S}}(k)^{\tau}\bar{\mathbb{S}}(k)\end{aligned} \tag{5.58}$$

Substituting $X(k+1)$ in Eq. (5.58), we may get

$$\begin{aligned}\Delta\mathcal{V}(k) = {} & [\bar{\Upsilon}_d\Gamma(\bar{E}X(k) + \bar{F}u(k) - 1_5 \otimes x_{r(k+1)} + \bar{d}_{ij})]^{\tau} * \\ & [\bar{\Upsilon}_d\Gamma(\bar{E}X(k) + \bar{F}u(k) - 1_5 \otimes x_{r(k+1)} + \bar{d}_{ij})] - \bar{\mathbb{S}}(k)^{\tau}\bar{\mathbb{S}}(k)\end{aligned} \tag{5.59}$$

Substituting the u_b(k) into Eq. (5.59), we may get

$$\begin{aligned}\Delta\mathcal{V}(k) = {} & [I_{18} \otimes (d(k) - d(0)) + \bar{\mathbb{S}}_d(k+1)]^{\tau} * \\ & [I_{18} \otimes (d(k) - d(0)) + \bar{\mathbb{S}}_d(k+1)] - \bar{\mathbb{S}}(k)^{\tau}\bar{\mathbb{S}}(k)\end{aligned} \tag{5.60}$$

Now denoting,

$$\bar{\beta} = [I_{18} \otimes (d(k) - d(0)) + \bar{\mathbb{S}}_d(k+1)]^{\tau} * [I_{18} \otimes (d(k) - d(0)) + \bar{\mathbb{S}}_d(k+1)] \tag{5.61}$$

$$\Delta\mathcal{V}(k) = \bar{\beta} - \bar{\mathbb{S}}(k)^{\tau}\bar{\mathbb{S}}(k) \tag{5.62}$$

For stability purpose the following conditions hold true:

$$0 \leq \bar{\beta} \leq \bar{\mathbb{S}}(k)^{\tau}\bar{\mathbb{S}}(k) \tag{5.63}$$

From Eq. (5.63), the global stability of DDSM formation protocol using Bartoszewicz's reaching law can be obtained by tuning the $\bar{\beta}$ so that the $\Delta\mathcal{V}(k) < 0$.

This completes the proof.

5.3.1 Simulation Results of DDSM Formation Protocol Using Bartoszewicz's Reaching Law

This Sect. 5.3.1 presents the comparison of simulation results of Quadcopter swarm with DDSM formation protocol using Bartoszewicz's reaching law. The simulation results of all agents for the X, Y, and Z positions, control efforts, and sliding variables using the formation controller designed using Bartoszewicz's reaching law are shown in Figs. (5.8, 5.9, 5.10, 5.11 and 5.12). Similarly, the tracking error using Bartoszewicz's reaching law is shown in Fig. 5.13.

5.4 Comparative Analysis of DDSOSM ST Protocol with DDSM Protocol Using Gao's Reaching Law

Section 5.4 gives the information about the comparison between the DDSOSM ST protocol and DDSM protocol using Gao's reaching law. The below formation protocol of DDSM using Gao's reaching law derived in Sect. 4.5 which is given below [6].

$$u_g(k) = -(\bar{\Upsilon}_d \Gamma \bar{F})^{-1}(\bar{\Upsilon}_d \Gamma \bar{E} X(k) + \bar{\Upsilon}_d \Gamma(-1_5 \otimes \tilde{E} x_{r(k)} + \bar{d}_{ij}) - (1 - \kappa_1 \tau)\bar{\mathbb{S}}(k) + \kappa_2 \tau sgn(\bar{\mathbb{S}}(k))) \quad (5.64)$$

The stability of the formation protocol using Gao's reaching law is also proved in Sect. 4.4.

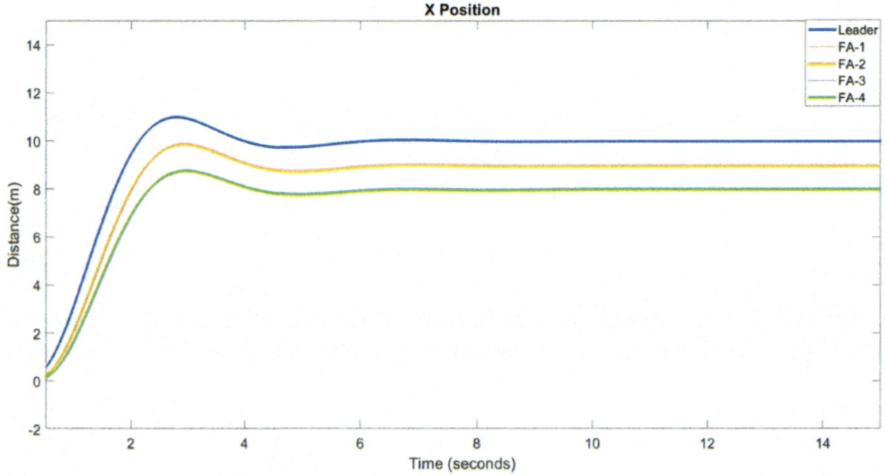

Fig. 5.8 X position of all agents using DDSM Bartoszewicz's reaching law

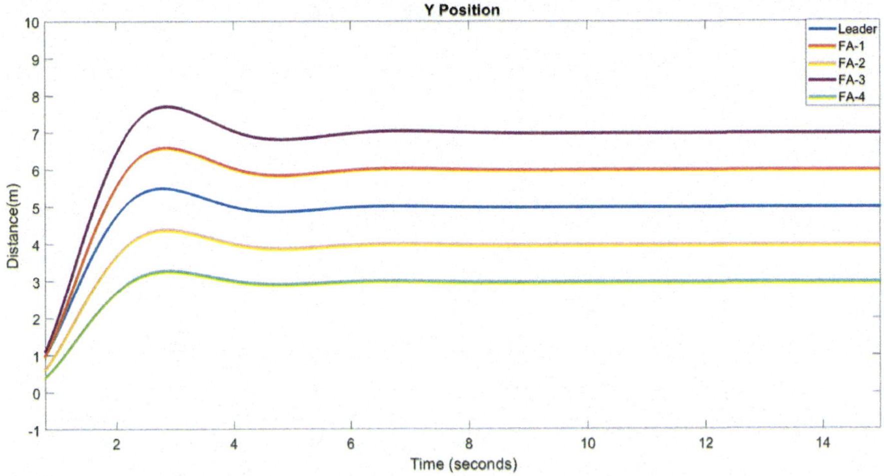

Fig. 5.9 Y position of all agents using DDSM Bartoszewicz's reaching law

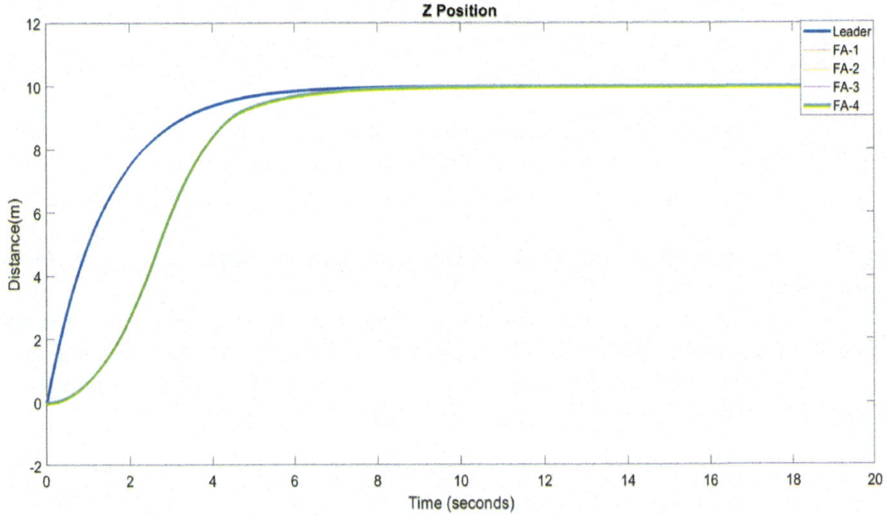

Fig. 5.10 Z position of all agents using DDSM Bartoszewicz's reaching law

5.4.1 Simulation Results of DDSM Formation Protocol Using Gao's Reaching Law

The simulation results of all agents for the X, Y, and Z positions, tracking error, and control effort using the formation controller designed using Gao's reaching law are shown in Figs. (5.14, 5.15, 5.16, 5.17 and 5.18). The control effort chattering band is –0.2 to +0.2 in the case of Gao's reaching law. Figure 5.19 shows the sliding variables

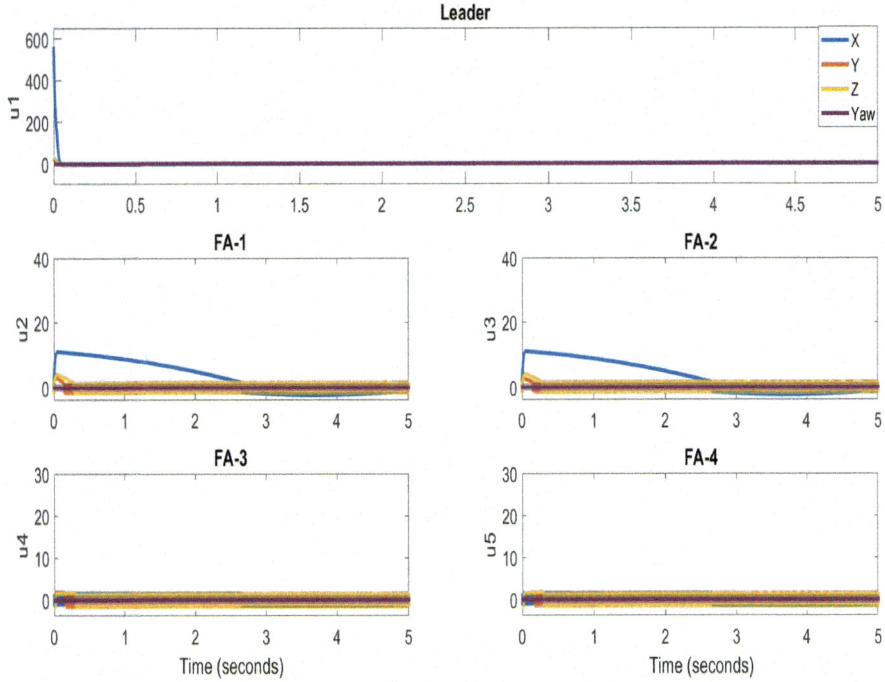

Fig. 5.11 Control effort of all agents using DDSM Bartoszewicz's reaching law

of all agents during the formation using Gao's reaching law. The quasi-sliding mode band during the formation is –0.01 to +0.01.

Further to evaluate the performance of formation, the tracking error and the control efforts using DDSOSM ST protocol are compared with two classical controllers Gao's reaching law and Bartoszewicz's reaching law which are quantified in Tables 5.1 and 5.2.

5.5 Conclusion

In this chapter, a DDSOSM ST protocol is used to design the formation of a Quadcopter swarm in a pattern. The DMAS concept is used to design the protocol. It is inferred from the simulation results that the proposed formation protocol brings all the Quadcopter in the given pattern in a finite time. The stability of the formation is also derived using the Lyapunov function. As the DST protocol is the second-order sliding mode control, it has the advantages of robustness, faster convergence, and less chattering than the first-order classical controllers. However the DST protocol may not be suitable for controlling systems with high-frequency dynamics. Fur-

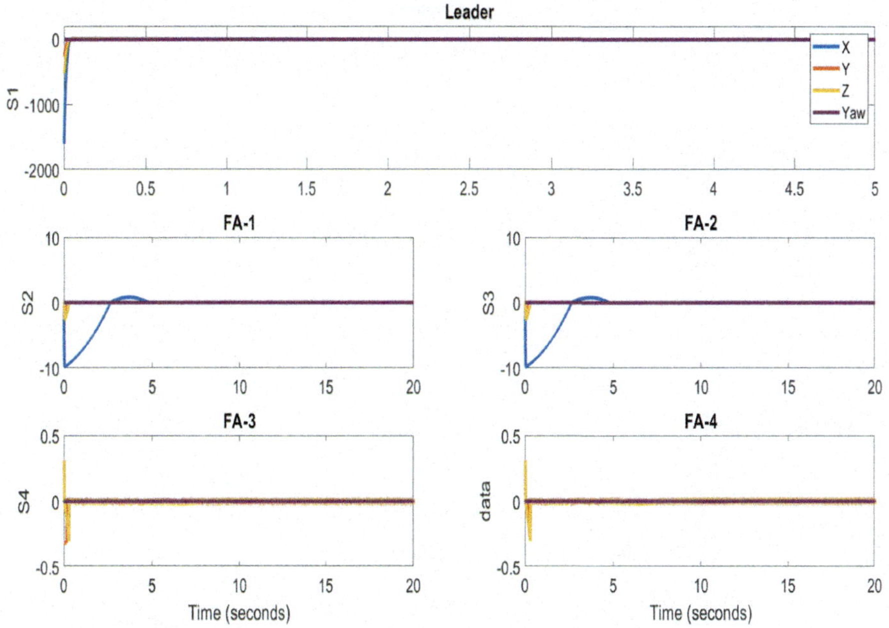

Fig. 5.12 Sliding surface of all agents using DDSM Bartoszewicz's reaching law

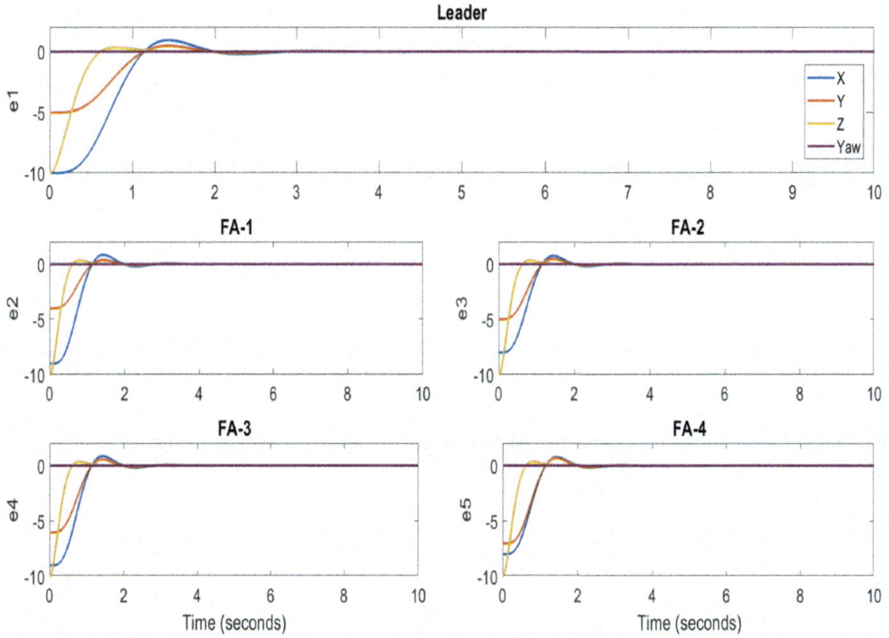

Fig. 5.13 Tracking error of all agents using DDSM Bartoszewicz's reaching law

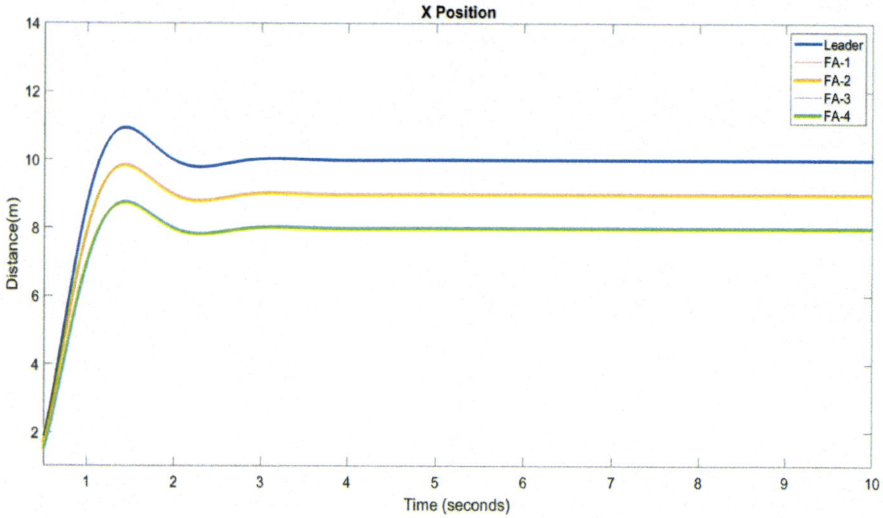

Fig. 5.14 X position of all agents using DDSM Gao's reaching law

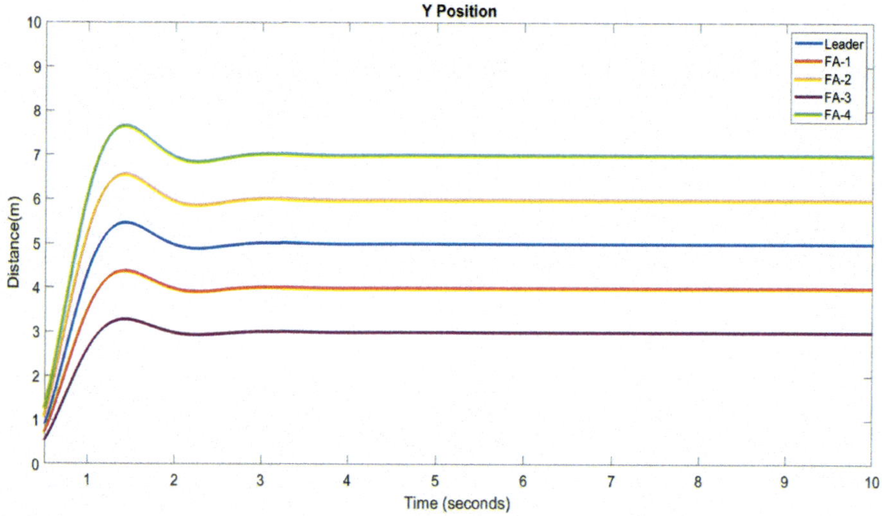

Fig. 5.15 Y position of all agents using DDSM Gao's reaching law

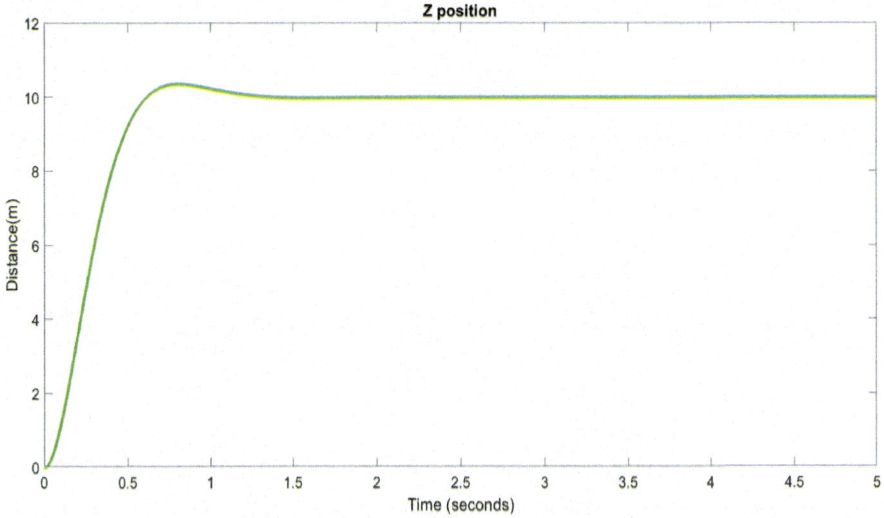

Fig. 5.16 Z position of all agents using DDSM Gao's reaching law

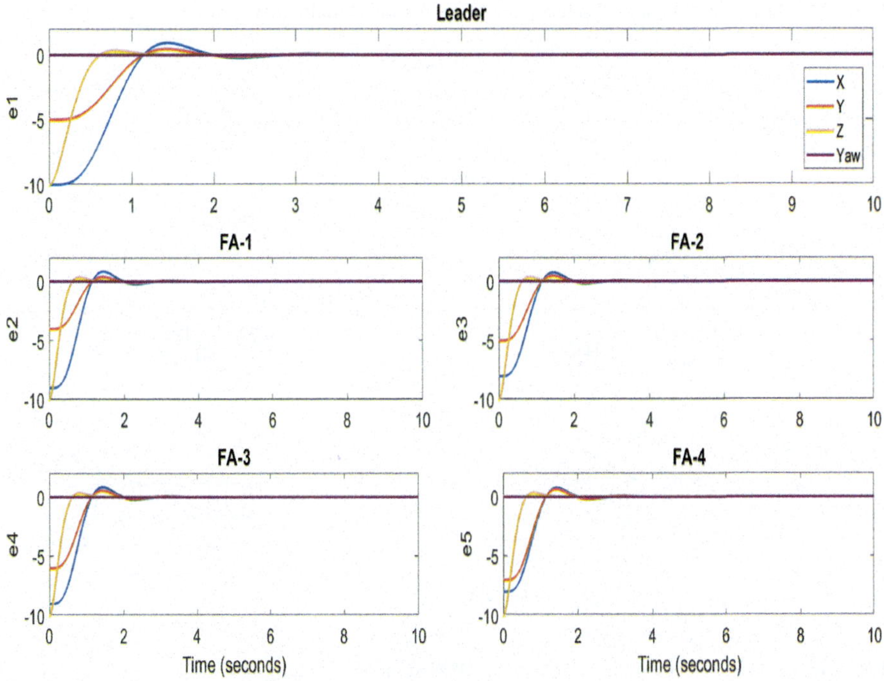

Fig. 5.17 Tracking error of all agents using DDSM Gao's reaching law

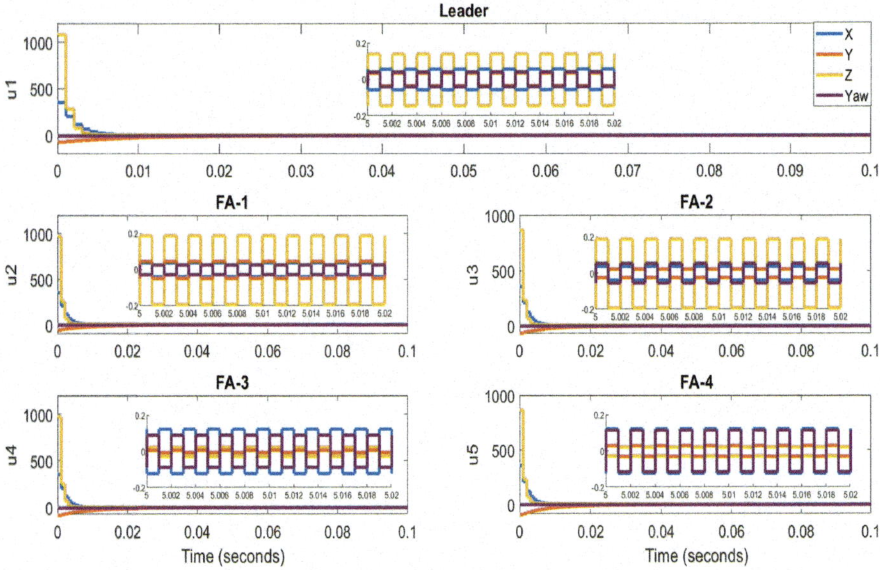

Fig. 5.18 Control effort of all agents using DDSM Gao's reaching law

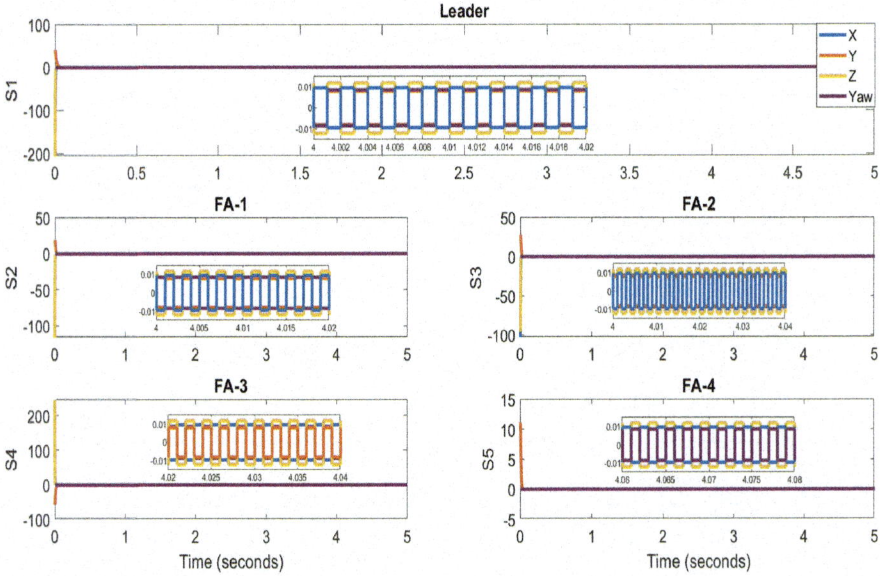

Fig. 5.19 Sliding surface of all agents using DDSM Gao's reaching law

Table 5.1 Comparison study of tracking error

Error [ISE]	Gao's protocol	Bartoszewicz's protocol	ST protocol
$e_{11}(X - axis)$	1.5×10^{-3}	3.2×10^{-4}	1.5×10^{-6}
$e_{12}(Y - axis)$	1.09×10^{-3}	2×10^{-4}	0.88×10^{-6}
$e_{13}(Z - axis)$	2.4×10^{-3}	1.4×10^{-4}	2.1×10^{-6}
$e_{14}(Yaw)$	1.2×10^{-3}	1.7×10^{-4}	1.05×10^{-6}
$e_{14}(Yaw)$	1.2×10^{-3}	1.7×10^{-4}	1.05×10^{-6}

Table 5.2 Comparison study of control effort

Control effort	Gao's protocol	Bartoszewicz's protocol	ST protocol
u_1	−0.2 to 0.2	0.0041	−0.001 to 0.001
u_2	−0.2 to 0.2	0.0032	−0.001 to 0.001
u_3	−0.2 to 0.2	0.0015	−0.001 to 0.001
u_4	−0.2 to 0.2	0.002	−0.001 to 0.001
u_5	−0.2 to 0.2	0.002	−0.001 to 0.001

ther, the performance of formation effort is compared with the other two classical designs using Gao's reaching law and Bartoszewicz's reaching law. The simulation results and the quantitative analysis show that the proposed formation control law outperforms the other controllers.

References

1. Salgado, I., Kamal, S., Chairez, I., Bandyopadhyay, B., Fridman, L.: Super-twisting-like algorithm in discrete time nonlinear systems. In: The 2011 International Conference on Advanced Mechatronic Systems, Zhengzhou, China, pp. 497–502 (2011)
2. Modi, A., Joshi, N., Mehta, A.: Robust discrete-time super twisting formation protocol for a 6-DOF Quadcopter swarm. ISA Trans. (2023). ISSN 0019-0578. https://doi.org/10.1016/j.isatra.2023.09.029
3. Salgado, I., Kamal, S., Bandyopadhyay, B., Chairez, I., Fridman, L.: Control of discrete time systems based on recurrent Super-Twisting-like algorithm. ISA Trans. **64**, 47–55 (2016). ISSN 0019-0578. https://doi.org/10.1016/j.isatra.2016.04.024
4. Patel, K., Mehta, A.: Discrete higher order sliding mode protocol for leader-following consensus of heterogeneous discrete multi-agent system. In: Lecture Notes in Electrical Engineering, pp. 1–10. Springer, Singapore (2019)
5. Bartoszewicz, A.: Discrete-time quasi-sliding-mode control strategies. IEEE Trans. Ind. Electron. **45**(4), 633–637 (1998). https://doi.org/10.1109/41.704892. Aug.
6. Gao, W., Wang, Y., Homaifa, A.: Discrete-time variable structure control systems. IEEE Trans. Ind. Electron. **42**(2), 117–122 (1995). https://doi.org/10.1109/41.370376

Chapter 6
Discrete-Time Sliding Mode Protocol Using Exponential Reaching Law for Formation of Quadcopter Swarm

6.1 Introduction to Discrete-Time Exponential Sliding Model Protocol

An exponential reaching law in sliding mode control (SMC) is a specific design approach that aims to smooth the switching behavior of traditional SMC, reducing chattering and improving control performance [1]. The primary advantage of exponential reaching law is its effectiveness in reducing chattering. The use of exponential functions leads to a smoother control signal, which in turn results in less high-frequency oscillation. Depending on the specific design and implementation, exponential reaching law can provide exponential stability. This is characterized by an exponentially decaying deviation from the sliding surface, ensuring a stable and controlled behavior.

Like other sliding mode control techniques, exponential reaching law is also inherently robust to uncertainties and disturbances, which is a crucial characteristic for many real-world applications.

The sliding surface for the discrete-time domain can be defined as

$$\mathbb{S}(k) = \Upsilon_d X(k) \tag{6.1}$$

where Υ_d is sliding mode gain which can be calculated using the Pole Placement method or LQR approach.

$$\mathbb{S}(k+1) = \Upsilon_d X(k+1) \tag{6.2}$$

$$\mathbb{S}(k+1) = \Upsilon_d (E(k)X(k) + F(k)u(k)) \tag{6.3}$$

The exponential type of reaching law is given as below [2]

A. Mehta and A. Modi, *Robust Sliding Mode Protocols for Formation of Quadcopter Swarm*, Studies in Systems, Decision and Control 521,
https://doi.org/10.1007/978-981-99-9726-8_6

$$\mathbb{S}(k+1) = (1 - q\tau)\varrho(k)\mathbb{S}(k) - \frac{\beta}{\varrho(k)} sgn(\mathbb{S}(k)) + \zeta(k) \tag{6.4}$$

where $q > 0$ and $0 < 1 - q\tau < 1$ are convergence parameter where τ shows the sampling time. $\beta > 0$ is a switching gain. Here, $q = [q_1, q_2, ..., q_N]^\tau \in R^{nN} > 0$, $\beta = [\beta_1, \beta_2, ..., \beta_N]^\tau \in R^{nN} > 0$, $\zeta(k) = [\zeta_1(k), \zeta_2(k), ..., \zeta_N(k)]^\tau \in R^{mN}$, $\varrho(k) = [\varrho_1(k), \varrho_2(k), ..., \varrho_N(k)]^\tau \in R^{nN}$, and $\delta < \varrho(k) < 1$.

$$\zeta(k) = \Upsilon_d F[d(k) - 2d(k-1) + d(k-2)] \tag{6.5}$$

$$\varrho(k) = \delta + (1 - \delta)exp^{-\theta|\mathbb{S}_i(k)|^\gamma} \tag{6.6}$$

is an exponential term that dynamically adapts the variations of the switching function, where $\theta > 0$, $\gamma > 0$; $\gamma \in N$, and $0 < \delta < 1$.

Comparing Eq. (6.3) with Eq. (6.4), we may get

$$\begin{aligned} u_e(k) = [-(\Upsilon_d \Gamma F)^{-1}[\Upsilon_d \Gamma E X(k) + \Upsilon_d \Gamma(-1_5 \otimes \tilde{E} x_0(k)) \\ - (1 - q\tau)\varrho(k)\mathbb{S}(k) + \frac{\beta}{\varrho(k)} sgn(\mathbb{S}(k)) - \zeta(k)]] \end{aligned} \tag{6.7}$$

Equation (6.7) is the control law of the DDSM protocol using exponential reaching law.

6.2 DDSM Formation Protocol for Quadcopter Swarm Using Exponential Reaching Law

To design the DDSM formation protocol using exponential reaching law, the pattern of the five Quadcopters as shown in Fig. 6.1 is considered.

As per the formation pattern, the position of the leader agent is considered as the (0,0,0) for the (X,Y,Z) position. The immediate follower-1 (FA-1) and follower-2 (FA-2) have the distance d_{ij} with leader agent in (X,Y,Z) position as $(-2, -1, 0)$ and $(2, -1, 0)$. Then the next follower-3 (FA-3) and follower-4 (FA-4) have the distance d_{ij} with leader agent in (X,Y,Z) position as $(-1, -3, 0)$ and $(1, -3, 0)$. The altitude of all the Quadcopters will remain the same or it can be presented as $d_{ij} = 0$ for all Quadcopters in Z position.

Consider the DMAS given in the Sect. 4.4.1 and define the local neighborhood formation error for the ith agent as

$$\delta(k) = \sum [a_{ij}[X_i(k) - X_j(k)] + \mathbb{P}_n[X_i(k) - x_r(k)] + d_{ij}] \tag{6.8}$$

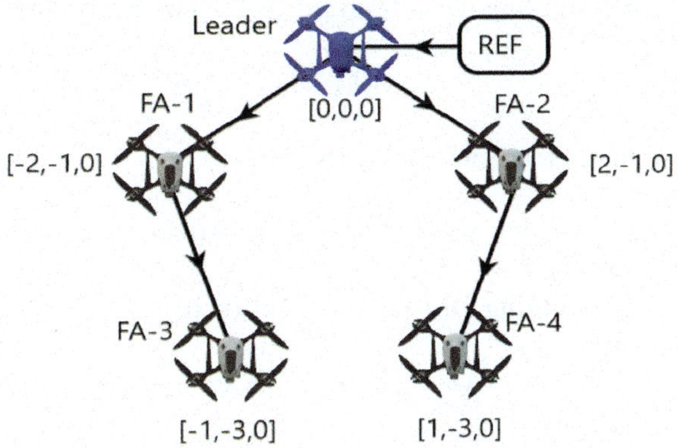

Fig. 6.1 Communication Topology

where $x_r(k)$ is the reference trajectory given to the leader and d_{ij} is the distance between the agents.

From Eq. (6.8), the global error can be written as

$$\mathbb{E}(k) = (((I_5 + \bar{\mathbb{D}} + \bar{\mathbb{B}})^{-1})(\bar{\mathbb{L}} + \bar{\mathbb{B}}) \otimes I_{18})\hat{x} \tag{6.9}$$

Further expanding Eq. (4.32), we may get

$$\mathbb{E}(k) = (((I_5 + \bar{\mathbb{D}} + \bar{\mathbb{B}})^{-1})(\bar{\mathbb{L}} + \bar{\mathbb{B}}) \otimes I_{18})(X(k) - 1_5 \otimes x_{r(k)} + \bar{d}_{ij}) \tag{6.10}$$

where $\hat{x} = X(k) - 1_5 \otimes x_{r(k)} + \bar{d}_{ij}$ and $(I_5 + \bar{\mathbb{D}} + \bar{\mathbb{B}})^{-1}$ must be non-singular matrix.

Equation (6.10) can be rewritten as

$$\mathbb{E}(k) = (\zeta \otimes I_{18})(X(k) - 1_5 \otimes x_{r(k)} + \bar{d}_{ij}) \tag{6.11}$$

where $\zeta = (I_5 + \bar{\mathbb{D}} + \bar{\mathbb{B}})^{-1})(\bar{\mathbb{L}} + \bar{\mathbb{B}})$

$$\mathbb{E}(k) = \Gamma(X(k) - 1_5 \otimes x_{r(k)} + \bar{d}_{ij}) \tag{6.12}$$

where $\Gamma = \zeta \otimes I_{18}$.

Now, define the sliding surface for the ith agent as

$$\mathbb{S}_i(k) = \Upsilon_d \delta(k) \tag{6.13}$$

Then Eq. (6.13) can be further written in the global form as

$$\bar{\mathbb{S}}(k) = \bar{\Upsilon}_d \mathbb{E}(k) \tag{6.14}$$

where $\bar{\Upsilon}_d = I_5 \otimes \Upsilon_d$,

Substituting the $\mathbb{E}$ from Eq. (6.12) into Eq. (6.14), we may get

$$\bar{\mathbb{S}}(k) = \bar{\Upsilon}_d \Gamma(X(k) - 1_5 \otimes x_{r(k)} + \bar{d}_{ij}) \tag{6.15}$$

Further, advancing Eq. (6.15)

$$\bar{\mathbb{S}}(k+1) = \bar{\Upsilon}_d \Gamma(X(k+1) - 1_5 \otimes x_{r(k+1)} + \bar{d}_{ij}) \tag{6.16}$$

Substituting Eq. (4.24) into Eq. (6.16)

$$\bar{\mathbb{S}}(k+1) = \bar{\Upsilon}_d \Gamma(\bar{E}X(k) + \bar{F}u(k) - 1_5 \otimes x_{r(k)} + \bar{d}_{ij}) \tag{6.17}$$

Now to derive the DDSM formation protocol, let us use the exponential reaching law for a discrete-time system defined for ith agent as

$$\mathbb{S}_i(k+1) = (1 - q_i \tau_i)\varrho_i(k)\mathbb{S}_i(k) - \frac{\beta_i}{\varrho_i(k)} sgn(\mathbb{S}_i(k)) + \zeta_i(k) \tag{6.18}$$

where $q_i > 0$ and $0 < 1 - q_i\tau_i < 1$ are convergence parameter where τ_i shows the sampling time of $i thagent.$ $\beta_i > 0$ is a switching gain. Here, $q_i = [q_1, q_2, ..., q_N]^\tau \in R^{nN} > 0$, $\beta_\rangle = [\beta_1, \beta_2, ..., \beta_N]^\tau \in R^{nN} > 0$, $\zeta_i(k) = [\zeta_1(k), \zeta_2(k), ..., \zeta_N(k)]^\tau \in R^{mN}$, $\varrho_i(k) = [\varrho_1(k), \varrho_2(k), ..., \varrho_N(k)]^\tau \in R^{nN}$, and $\delta_i < \varrho_i(k) < 1$.

$$\zeta_i(k) = \Upsilon_d F[d(k) - 2d(k-1) + d(k-2)] \tag{6.19}$$

$$\varrho_i(k) = \delta_i + (1 - \delta_i) exp^{-\theta_i |\mathbb{S}_i(k)|^{\gamma_i}} \tag{6.20}$$

is an exponential term that dynamically adapts the variations of the switching function, where $\theta_i > 0$, $\gamma_i > 0$; $\gamma_i \in N$, and $0 < \delta_i < 1$.

Using Eq. (6.18), we may write the exponential reaching law in global form as

$$\bar{\mathbb{S}}(k+1) = (1 - \bar{q}\tau)\bar{\varrho}(k)\bar{\mathbb{S}}(k) - \frac{\bar{\beta}}{\bar{\varrho}(k)} sgn(\bar{\mathbb{S}}(k)) + \bar{\zeta}(k) \tag{6.21}$$

where $\bar{q} = I_5 \otimes q_i$, $\bar{\varrho} = I_5 \otimes \varrho_i$, $\bar{\beta} = I_5 \otimes \beta$, $\bar{\zeta}(k) = I_5 \otimes \zeta(k)$, and

$$\bar{\zeta}(k) = \bar{\Upsilon}_d \bar{F}[\bar{d}(k) - 2\bar{d}(k-1) + \bar{d}(k-2)] \tag{6.22}$$

$$\bar{\varrho}(k) = \bar{\delta} + (1 - \bar{\delta}) exp^{-\bar{\theta}|\bar{\mathbb{S}}(k)|^{\bar{\gamma}}} \tag{6.23}$$

where $\bar{\theta} = I_5 \otimes \theta_i$, $\bar{\gamma} = I_5 \otimes \gamma_i$, and $0 < \bar{\delta} < 1$.

Now equating Eqs. (6.17) and (6.21), we may get the DDSM formation protocol using the exponential reaching law for the Quadcopter swarm.

$$u_e(k) = -(\bar{\Upsilon}_d \Gamma \bar{F})^{-1}(\bar{\Upsilon}_d \Gamma \bar{E} X(k) + \bar{\Upsilon}_d \Gamma(-1_5 \otimes \tilde{E} x_{r(k)} + \bar{d}_{ij}) \\ - (1 - \bar{q}\tau)\bar{\varrho}(k)\bar{\mathbb{S}}(k) - \frac{\bar{\beta}}{\bar{\varrho}(k)} sgn(\bar{\mathbb{S}}(k)) + \bar{\zeta}(k) \quad (6.24)$$

Remark 1 The disturbance vector $d(k)$ will have the boundedness with its further rate of change as

$$\zeta_i(k) = \Upsilon_d F[d(k) - d(k-1)] \\ ||\zeta_i(k)|| \leq \rho \quad (6.25)$$

where ρ is the upper bound of vector $\zeta_i(k)$. From this remark, the $d(k)$ possesses the below lemma as

Lemma 1 $d(k) = O(T)$, $d(k) - d(k-1) = O(T^2)$, *and* $d(k) - d(k-1) - d(k-2) = O(T^3)$.

From Eq. (6.25) and Lemma 1, it can be stated that the maximum deviation after the formation achieved that is measured in terms of width of QSMB of sliding variable with the protocol given in Eq. (6.24) will remain in the order $O(T^3)$.

6.2.1 *Analysis of Number of Steps Required for Formation of Quadcopter Swarm with DDSM Using Exponential Reaching Law*

Further, it is required to estimate the number of steps required for achieving the formation among all the Quadcopter agents for the DMAS which is derived in Theorem 6.2.

Theorem 6.1 *In homogeneous DMAS as defined in Eq. (4.24), the number of steps requires to reach the global formation among all agents using the derived DDSM protocol using exponential reaching law given in Eq. (6.24) is max(η+1), where*

$$\eta = log_{(1-q_i\tau_i)} \frac{\psi_3 q_i \tau_i \sigma_i + \psi_2(\beta_i - \zeta\psi_3)}{\psi_1\psi_3 q_i \tau_i |s_i(0)| + \psi_2(\beta_i - \zeta\psi_3)} \quad (6.26)$$

is the positive real number.

Proof To prove Theorem 6.2, let us consider two conditions that are possible during the formation of ith agent as $\mathbb{S}_i(0) > 0$ and $\mathbb{S}_i(0) < 0$.

(i) If $\mathbb{S}_i(0) > 0$**,**

Let us consider the exponential reaching law for the ith agent as given in Eq. (6.18),

$$\mathbb{S}_i(k+1) = (1 - q_i\tau_i)\varrho_i(k)\mathbb{S}_i(k) - \frac{\beta_i}{\varrho_i(k)} sgn(\mathbb{S}_i(k)) + \zeta_i(k) \quad (6.27)$$

Let Eq. (6.27) is derived for K=0, 1, 2,... as

$$\mathbb{S}_i(1) = (1 - q_i\tau_i)\varrho_i(0)s_i(0) - \frac{\beta_i}{\varrho_i(0)} sgn(\mathbb{S}_i i(0)) + \zeta_i(0)$$

$$\mathbb{S}_i i(2) = (1 - q_i\tau_i)^2\varrho_i(1)\mathbb{S}_i(0) - (1 - q_i\tau_i)\frac{\beta_i}{\varrho_i(0)} + (1 - q_i\tau_i)\zeta_i(0) - \frac{\beta_i}{\varrho_i(1)} + \zeta_i(1)$$

$$\vdots$$

Likewise for $k = \Theta$, where Θ is a positive integer.

$$\begin{aligned} \mathbb{S}_i(\Theta + 1) = (1 - q_i\tau_i)^{\Theta+1}\mathbb{S}_i(0) \textstyle\prod_{k=0}^{\Theta} \varrho_i(k) - \sum_{p=0}^{\Theta-1} \\ (1 - q_i\tau_i)^{\Theta-p}[\varrho_i(\Theta)\varrho_i(\Theta - 1)...\varrho_i(p + 1)][\tfrac{\beta_i}{\varrho_i(p)} - \zeta_i(p)] \\ - \tfrac{\beta_i}{\varrho_i(\Theta)} + \zeta_i(\Theta) \end{aligned} \tag{6.28}$$

For the ease, define

$$\Psi_1 = \prod_{k=0}^{\Theta}$$

and

$$\Psi_2 = \sum_{p=0}^{\Theta-1}[\varrho_i(\Theta)\varrho_i(\Theta - 1)...\varrho_i(p + 1)]$$

Equation (6.28) can be rewritten as

$$\begin{aligned} \mathbb{S}_i(\mu + 1) = (1 - q_i\tau_i)^{\Theta+1}\mathbb{S}_i(0)\Psi_1 - \Psi_2 \sum_{p=0}^{\Theta-1} (1 - q_i\tau_i)^{\Theta-p} \\ [\tfrac{\beta_i}{\varrho_i(p)} - \zeta_i(p)] - \tfrac{\beta_i}{\varrho_i(\Theta)} + \zeta_i(\Theta) \end{aligned} \tag{6.29}$$

$$\mathbb{S}_i(\Theta + 1) = (1 - q_i\tau_i)^{\Theta+1}\mathbb{S}_i(0)\Psi_1 - \Psi_2 \sum_{p=0}^{\Theta}(1 - q_i\tau_i)^{\Theta-p}[\frac{\beta_i}{\varrho_i(p)} - \zeta_i(p)] \tag{6.30}$$

Let,

$$\Psi_3 = \varrho_i(p)$$

and $\zeta_i(p)$ is assumed to be bounded.

$$\mathbb{S}_i(\Theta+1) = (1-q_i\tau_i)^{\Theta+1}\mathbb{S}_i(0)\Psi_1 - \Psi_2[\frac{\beta_i}{\Psi_3} - \zeta_i]\sum_{p=0}^{\Theta}(1-q_i\tau_i)^{\Theta-p} \tag{6.31}$$

Comparing Eq. (6.31) with the geometric progression series $s_n = \frac{a(1-r^n)}{1-r}$ where a is the first term of the series, r is the common ration, and n is the total number of terms in the series, we may write Eq. (6.31) as

$$\mathbb{S}_i(\Theta+1) = (1-q_i\tau_i)^{\Theta+1}\Theta_i(0)\Psi_1 - \Psi_2[\frac{\beta_i}{\Psi_3} - \zeta_i][\frac{1-(1-q_i\tau_i)^{\Theta+1}}{q_i\tau_i}] \tag{6.32}$$

Let's define $\Theta + 1 = \eta$. Hence

$$\mathbb{S}_i(\eta) = (1-q_i\tau_i)^{\eta}\mathbb{S}_i(0)\Psi_1 - \Psi_2[\frac{\beta_i}{\Psi_3} - \zeta_i][\frac{1-(1-q_i\tau_i)^{\eta}}{q_i\tau_i}] \tag{6.33}$$

From Eq. (6.33), it can be inferred that for any $\eta^* \geq \eta$ step, the local neighborhood error of ith agent once reaches the sliding surface, it will move in a zigzag way called QSMB. Hence, Eq. (6.33) will be

$$\mathbb{S}_i(\eta) = (1-q_i\tau_i)^{\eta}\mathbb{S}_i(0)\Psi_1 - \Psi_2[\frac{\beta_i}{\Psi_3} - \zeta_i][\frac{1-(1-q_i\tau_i)^{\eta}}{q_i\tau_i}] \tag{6.34}$$

From Eq. (6.34), we can derive the required number of steps for reaching the formation as

$$\eta = log_{(1-q_i\tau_i)}\frac{\Psi_3 q_i\tau_i\mathbb{S}_i + \Psi_2(\beta_i - \zeta_i\Psi_3)}{\Psi_1\Psi_3 q_i\tau_i|\mathbb{S}_i(0)| + \Psi_2(\beta_i - \zeta_i\Psi_3)} \tag{6.35}$$

(ii) If $\mathbb{S}_i(0) < 0$,
In this case, the $\mathbb{S}_i(\eta+1)$ will become negative for all $\eta^* \geq \eta + 1$. So Eq. (6.35) will be

$$\mathbb{S}_i(\eta+1) \leq -(1-q_i\tau_i)^{\eta+1}\mathbb{S}_i(0)\Psi_1 + \Psi_2[\frac{\beta_i}{\Psi_3} - \zeta_i] [\frac{1-(1-q_i\tau_i)^{\eta+1}}{q_i\tau_i}] \tag{6.36}$$

This completes the proof.

Now, the global formation stability can be derived for the DMAS using the Lyapunov function from Theorem 6.1 as derived below.

Theorem 6.2 ([3]) *The global formation of DMAS as defined in Eq. (4.26) is said to be stable for derived DDSM protocol using exponential reaching law in Eq. (6.24) if the closed-loop error dynamics in Eq. (6.10) of DMAS drive toward the global sliding surface and maintain on it for any gain $\bar{q}, \bar{\theta}, \bar{\gamma}, \bar{\beta} > 0$ and $0 < \bar{\delta} < 1$, provided the following condition is true:*

$$0 \leq \alpha < \bar{\mathbb{S}}^{\tau}(k)\bar{\mathbb{S}}(k) \tag{6.37}$$

where

$$\alpha = [(1 - \bar{q}\tau)\bar{\varrho}(k)\bar{\mathbb{S}}(k) - \frac{\bar{\beta}}{\bar{\varrho}(k)} sgn(\mathbb{S}(k)) + \bar{\zeta}(k)]^{\tau} - [(1 - \bar{q}\tau)\bar{\varrho}(k)\bar{\mathbb{S}}(k) - \frac{\bar{\beta}}{\bar{\varrho}(k)} sgn(\bar{\mathbb{S}}(k)) + \bar{\zeta}(k)] \tag{6.38}$$

Proof Let the Lyapunov function be defined as

$$\mathcal{V}(k) = \bar{\mathbb{S}}^{\tau}(k)\bar{\mathbb{S}}(k) \tag{6.39}$$

The forward derivative function can be derived from Eq. (6.38) as

$$\Delta\mathcal{V}(k) = \mathcal{V}(k+1) - \mathcal{V}(k) \tag{6.40}$$

$$\Delta\mathcal{V}(k) = \bar{\mathbb{S}}^{\tau}(k+1)\bar{\mathbb{S}}(k+1) - \bar{\mathbb{S}}^{\tau}(k)\bar{\mathbb{S}}(k) \tag{6.41}$$

The DMAS is said to be stable if and only if $\Delta\mathcal{V}(k) < 0$.
Now, substituting Eq. (6.17) in Eq. (6.41),

$$\Delta\mathcal{V}(k) = [\bar{\Upsilon}_d \Gamma(\bar{E}X(k) + \bar{F}u(k) - 1_5 \otimes x_{r(k)} + \bar{d}_{ij})]^{\tau} [\bar{\Upsilon}_d \Gamma(\bar{E}X(k) + \bar{F}u(k) - 1_5 \otimes x_{r(k)} + \bar{d}_{ij})] - \bar{\mathbb{S}}^{\tau}(k)\bar{\mathbb{S}}(k) \tag{6.42}$$

Putting Eq. (6.24) into Eq. (6.42), we get

$$\Delta\mathcal{V}(k) = [(1 - \bar{q}\tau)\bar{\varrho}(k)\bar{\mathbb{S}}(k) - \frac{\bar{\beta}}{\bar{\varrho}(k)} sgn(\mathbb{S}(k)) + \bar{\zeta}(k)]^{\tau} - [(1 - \bar{q}\tau)\bar{\varrho}(k)\bar{\mathbb{S}}(k) - \frac{\bar{\beta}}{\bar{\varrho}(k)} sgn(\mathbb{S}(k)) + \bar{\zeta}(k)] - \bar{\mathbb{S}}^{\tau}(k)\bar{\mathbb{S}}(k) \tag{6.43}$$

denoting,

$$\alpha = [(1-\bar{q}\tau)\bar{\varrho}(k)\bar{\mathbb{S}}(k) - \frac{\bar{\beta}}{\bar{\varrho}(k)}sgn(\mathbb{S}(k)) + \bar{\zeta}(k)]^{\tau} - [(1-\bar{q}\tau)\bar{\varrho}(k)\bar{\mathbb{S}}(k) - \frac{\bar{\beta}}{\bar{\varrho}(k)}sgn(\mathbb{S}(k)) + \bar{\zeta}(k)] \quad (6.44)$$

$$\Delta\mathcal{V}(k) = \alpha - \bar{\mathbb{S}}^{\tau}(k)\bar{\mathbb{S}}(k) \quad (6.45)$$

For stability purpose the following conditions hold true:

$$0 \leq \alpha < \bar{\mathbb{S}}^{\tau}(k)\bar{\mathbb{S}}(k) \quad (6.46)$$

From Eq. (6.46), $\Delta\mathcal{V}(k) < 0$ can be achieved by tuning ν nearly equal to zero with the proper selection of $\bar{q}, \bar{\theta}, \bar{\gamma}, \bar{\beta}, \bar{\delta} > 0$. This completes the proof.

Remark 2 The DMAS is said to be stable since the defined global exponential term $\bar{\varrho}(k)$ is always positive. When the initial states of the DMAS are far from the defined global sliding surface $\bar{\mathbb{S}}(k)$, the $\bar{\varrho}(k)$ tends to remain on the defined $\bar{\delta}$ that in turn shows the term $(1-\bar{q}\tau)\bar{\varrho}(k)$ becomes $(1-\bar{q}\tau)\bar{\delta}$, which is smaller than $(1-\bar{q}\tau)$ and $\frac{\bar{\beta}}{\bar{\varrho}(k)}$ becomes $\frac{\bar{\beta}}{\bar{\delta}}$, which is greater than $\bar{\beta}$. This condition provides fast convergence toward the global sliding surface and improvement in the reaching time. When the states reach near to the surface, the exponential term $\bar{\varrho}(k)$ will become one that in turn shows the terms $(1-\bar{q}\tau)\bar{\varrho}(k)$ and $\frac{\bar{\beta}}{\bar{\varrho}(k)}$ to become $(1-\bar{q}\tau)$ and $\bar{\beta}$, respectively, which provides smaller control gains to remain in formation. Moreover, due to the adaption capability of the defined exponential term $\bar{\varrho}(k)$ around the switching function $\bar{\beta}$, the width of the QSMB will be reduced with magnitude in the order of $O(\tau^3)$ and this gradual increment of the exponential term helps to limit the developed chattering around the global sliding surface.

6.2.2 *Simulation Results of DDSM Formation Protocol Using Exponential Reaching Law*

The simulation results are carried out on a swarm of five homogeneous Quadcopters with the formation topology as shown in Fig. 6.3. One of the Quadcopters acts as leader and others as followers. The leader Quadcopter has access to the reference trajectory. The Laplacian $\mathbb{L}$, adjacency $\mathbb{A}$, degree $\mathbb{D}$, and diagonal matrices $\mathbb{B}$ are derived as

$$\mathbb{L} = \begin{bmatrix} 2 & 0 & 0 & 0 & 0 \\ -1 & 2 & 0 & 0 & 0 \\ -1 & 0 & 2 & 0 & 0 \\ 0 & -1 & 0 & -1 & 0 \\ 0 & 0 & -1 & 0 & 1 \end{bmatrix} \quad (6.47)$$

$$\mathbb{D} = \begin{bmatrix} 2\,0\,0\,0\,0 \\ 0\,2\,0\,0\,0 \\ 0\,0\,2\,0\,0 \\ 0\,0\,0\,1\,0 \\ 0\,0\,0\,0\,1 \end{bmatrix} \tag{6.48}$$

$$\mathbb{P} = \begin{bmatrix} 1\,0\,0\,0\,0 \\ 0\,0\,0\,0\,0 \\ 0\,0\,0\,0\,0 \\ 0\,0\,0\,0\,0 \\ 0\,0\,0\,0\,0 \end{bmatrix} \tag{6.49}$$

$$\mathbb{A} = \begin{bmatrix} 0\,0\,0\,0\,0 \\ 1\,0\,0\,0\,0 \\ 1\,0\,0\,0\,0 \\ 0\,1\,0\,0\,0 \\ 0\,0\,1\,0\,1 \end{bmatrix} \tag{6.50}$$

The sliding mode gain for the designed DDSM protocol using exponential reaching law is calculated using the LQR approach. The gain for each agent κ_1 and κ_2 is chosen as [1 1 1 1] and [2 4 3 1], respectively. The value of η_i is 0.5 for the $i^t h$ agent. Figure 6.2 shows the X-axis position trajectory of five Quadcopter agents, where the X position of the leader is 10 m from the origin. As the d_{ij} for FA-1 and FA-2 from the leader is -2 and 2 m, the X position of FA-1 and FA-2 is 8 m and 10m from the origin. Similarly, X position of FA-3 and FA-4 is 9 and 11 m. Figure 6.3 shows the Y position of five Quadcopters. The Y position of the leader is 5 m from the origin. As the d_{ij} for FA-1 and FA-2 is -1 m, so that the Y position of FA-1 and FA-2 is 4 m. Similarly, the Y position for the FA-2 and FA-3 is 2 m. The d_{ij} is zero for the Z position, altitude of all agents is the same at 10 m shown in Fig. 6.4. Figure 6.5 shows the tracking error of the X, Y, Z, and Yaw positions for all the agents. It can be noted that the error for all the agents reaches to zero in a finite time. Figure 6.6 shows the control efforts required to control the desired location generated by the individual agent. After settling at the desired location, maximum control effort generated by the leader agent varies in the band of -0.00003 to 0.00003. The chattering in control effort is shown as a magnified window in Fig. 6.6. Figure 6.7 shows the sliding surface of each Quadcopter. The sliding surface band is -2.2×10^{-6} to $+2.2 \times 10^{-6}$.

Hence to check the robustness of formation for Quadcopter swarm against the external disturbance given to the plant. Even if there is a variation in mass while using for the payload, still the DDSM protocol using exponential reaching law will provide the better results.

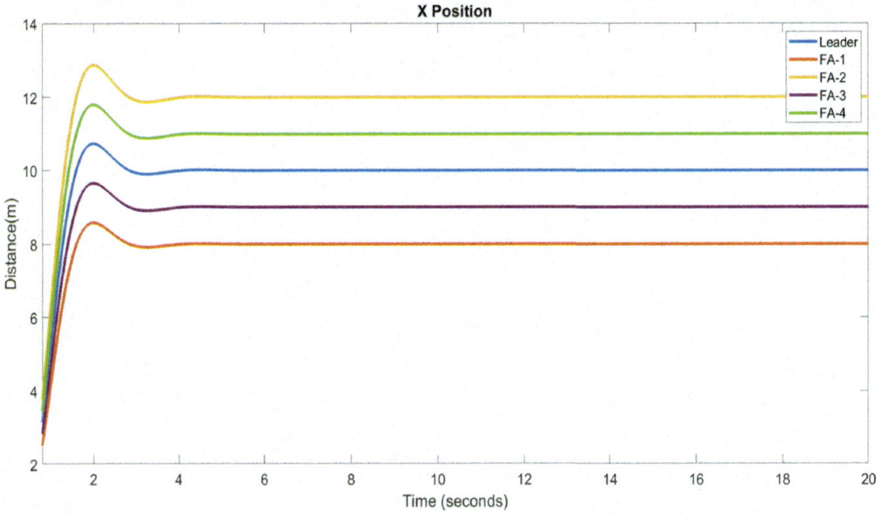

Fig. 6.2 X position of all agents using DDSM exponential reaching law

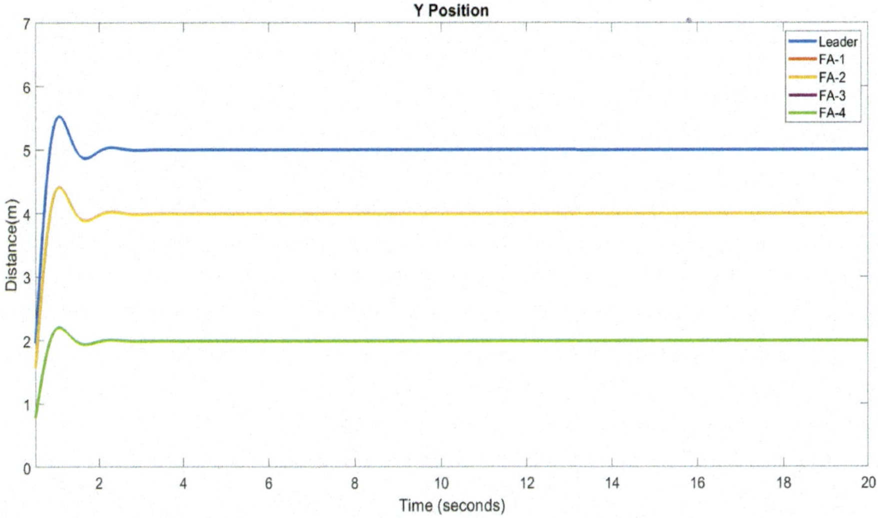

Fig. 6.3 Y position of all agents using DDSM exponential reaching law

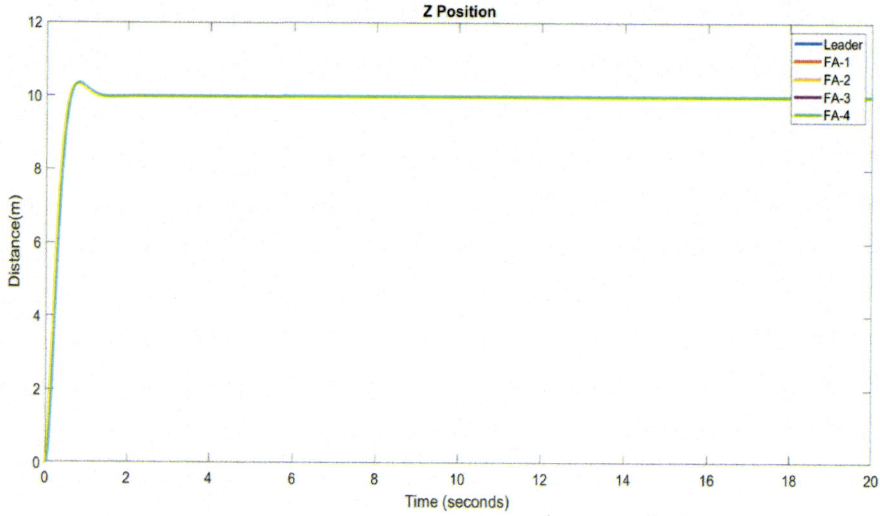

Fig. 6.4 Z position of all agents using DDSM exponential reaching law

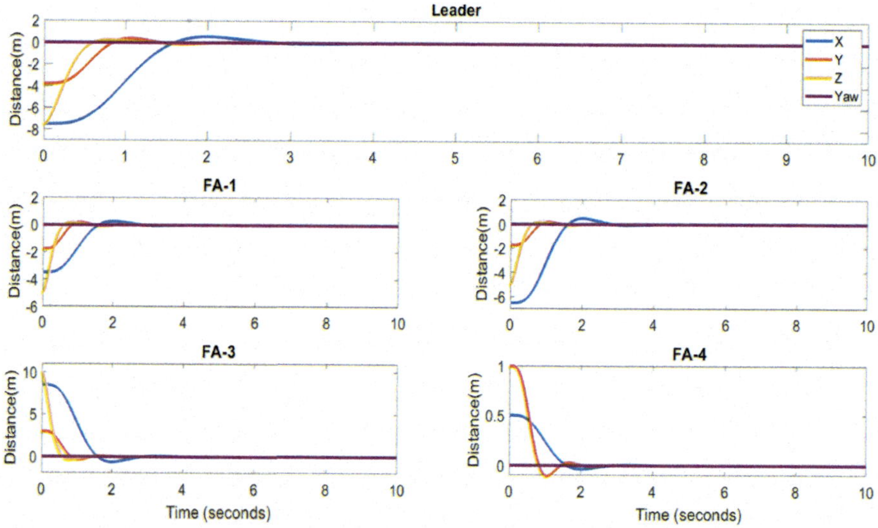

Fig. 6.5 Tracking error of all agents using DDSM exponential reaching law

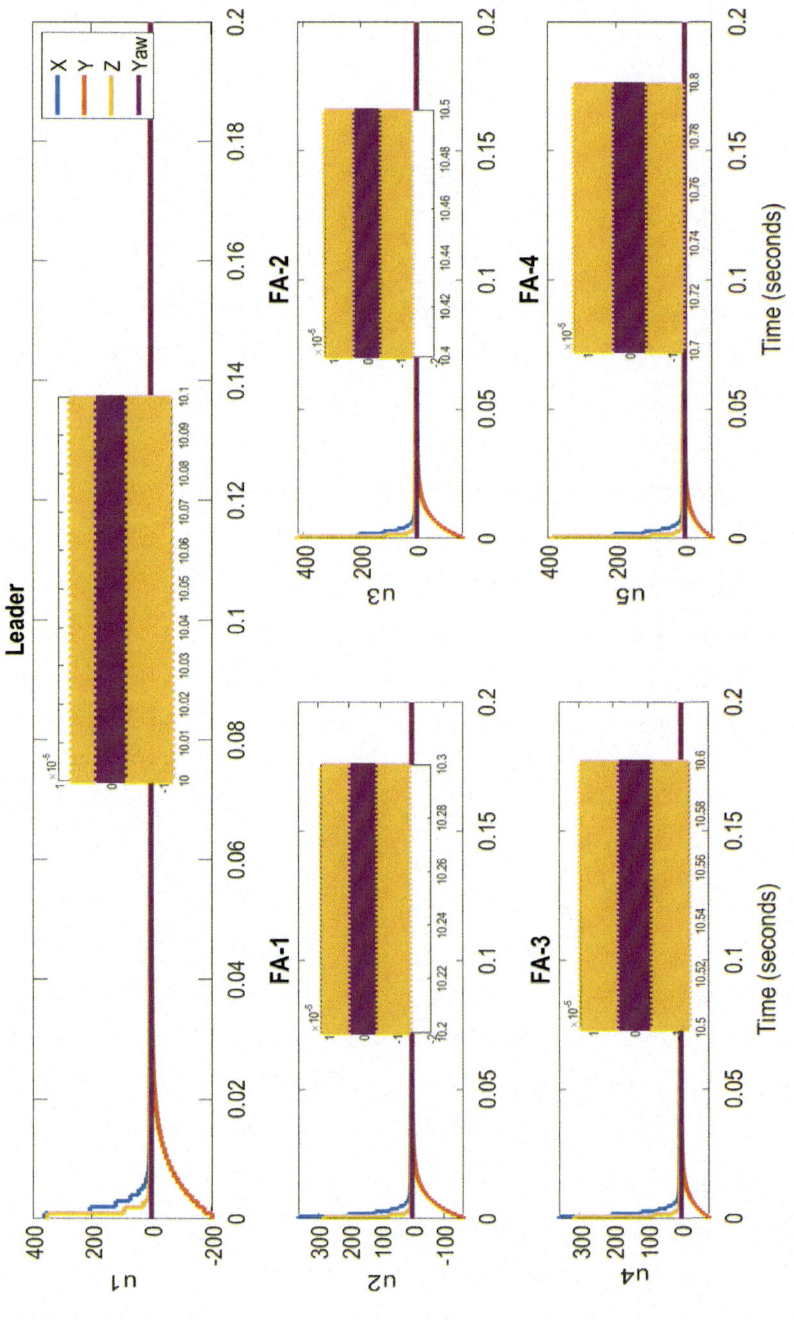

Fig. 6.6 Control efforts of all agents using DDSM exponential reaching law

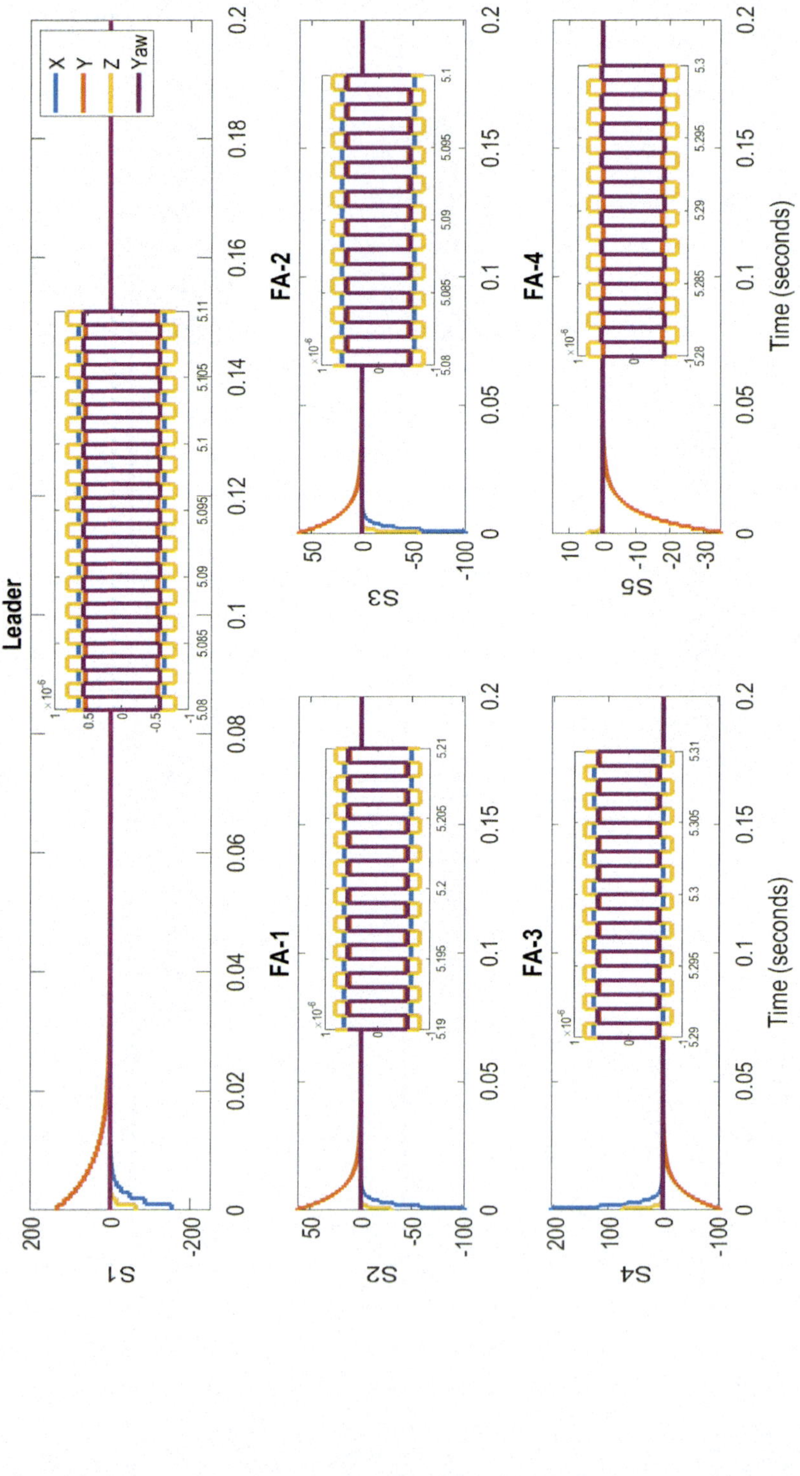

Fig. 6.7 Sliding surface of all agents using DDSM exponential reaching law

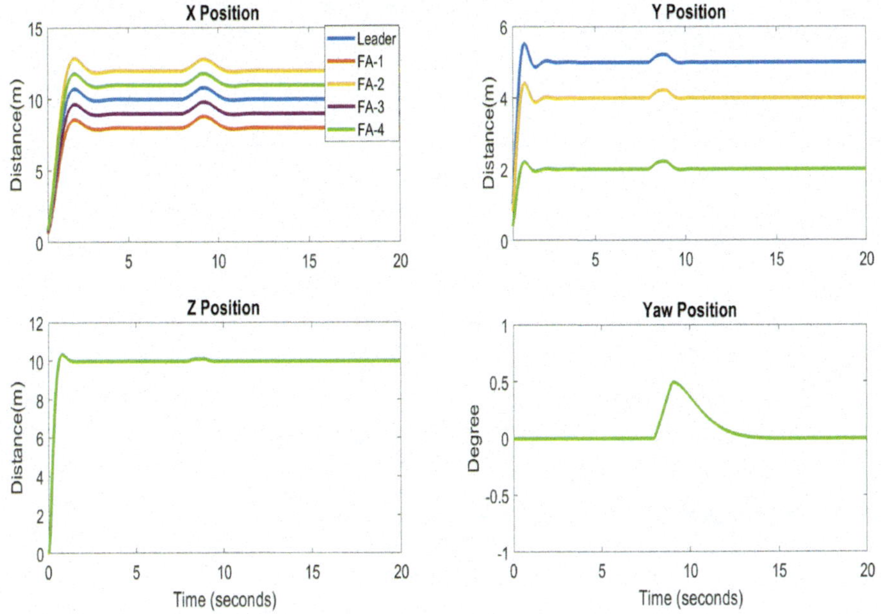

Fig. 6.8 Results of all agents under the disturbance

6.3 Comparative Analysis of DDSM Using Exponential Reaching Law with DDSM Protocol Using Gao's Reaching Law

To show the efficacy of the DDSM protocol using exponential reaching law, the simulation results are compared with the other classical DDSM protocol designed using Gao's reaching law [4] given in Eq. (4.88).

$$u_g(k) = -(\bar{\Upsilon}_d \Gamma \bar{F})^{-1}(\bar{\Upsilon}_d \Gamma \bar{E} X(k) + \bar{\Upsilon}_d \Gamma(-1_5 \otimes \tilde{E} x_{r(k)} + \bar{d}_{ij}) - (1 - \kappa_1 \tau)\bar{\mathbb{S}}(k) + \kappa_2 \tau sgn(\bar{\mathbb{S}}(k))) \quad (6.51)$$

The simulation results of all agents for the X, Y, and Z positions are shown in Figs. 6.9, 6.10 and 6.11. Figure 6.12 shows the tracking error of all agents. Control efforts and sliding variables using the formation controller using Gao's reaching law are given in Figs. 6.13 and 6.14. The chattering band of control effort and sliding surface is −0.002 to +0.002 and −0.0002 to +0.0002 in case Gao's reaching law.

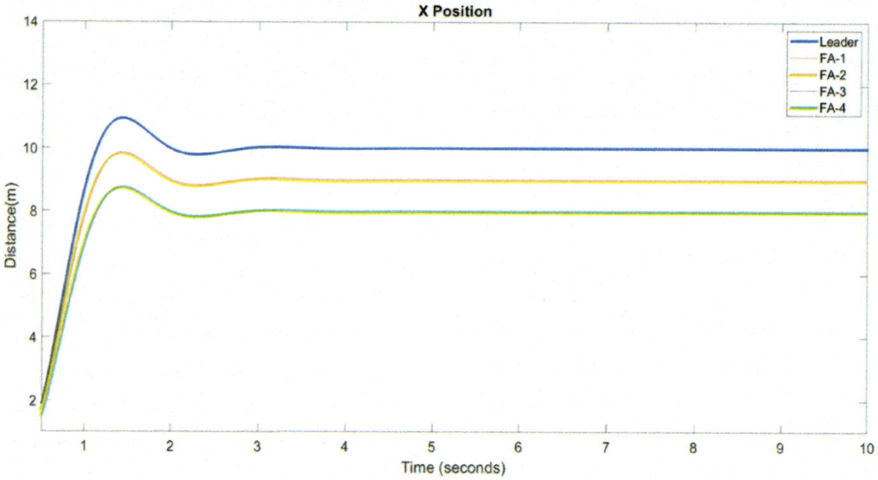

Fig. 6.9 X position of all agents using DDSM Gao's reaching law

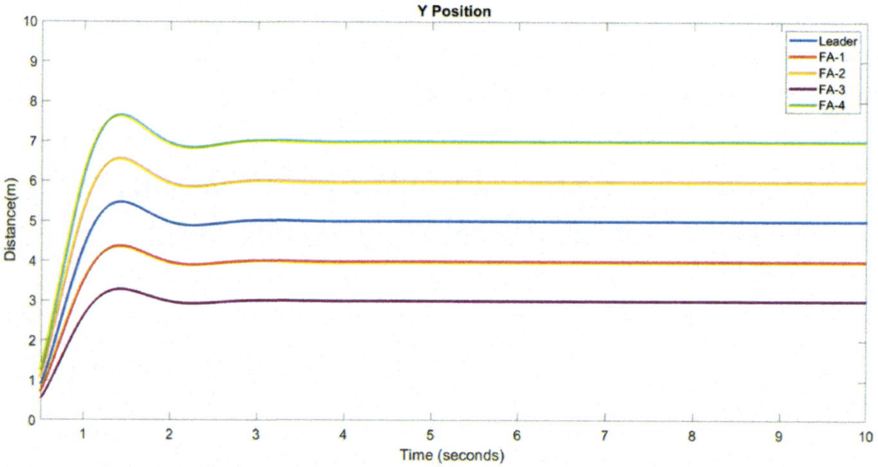

Fig. 6.10 Y position of all agents using DDSM Gao's reaching law

Also to evaluate the effectiveness of formation, the sliding surfaces and control efforts using exponential and Gao's reaching law are quantified as given in Tables 6.1 and 6.2. It is observed that the formation achieved DDSM protocol using exponential reaching law is faster than the DDSM protocol designed using Gao's reaching law.

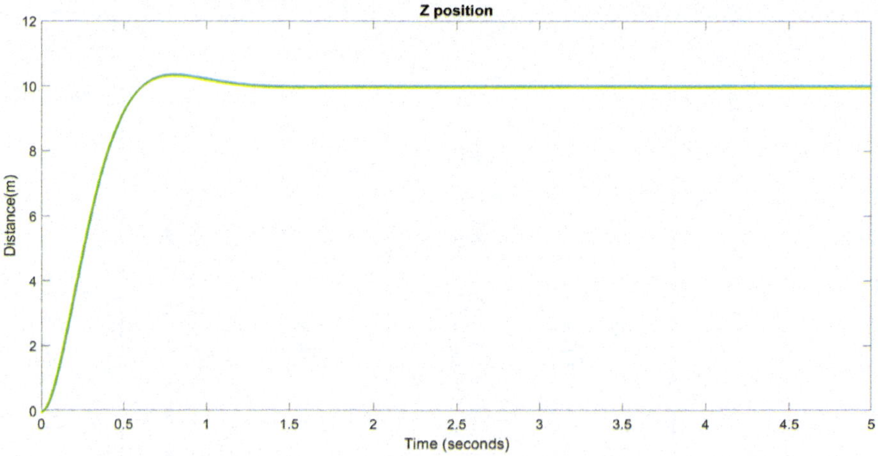

Fig. 6.11 Z position of all agents using DDSM Gao's reaching law

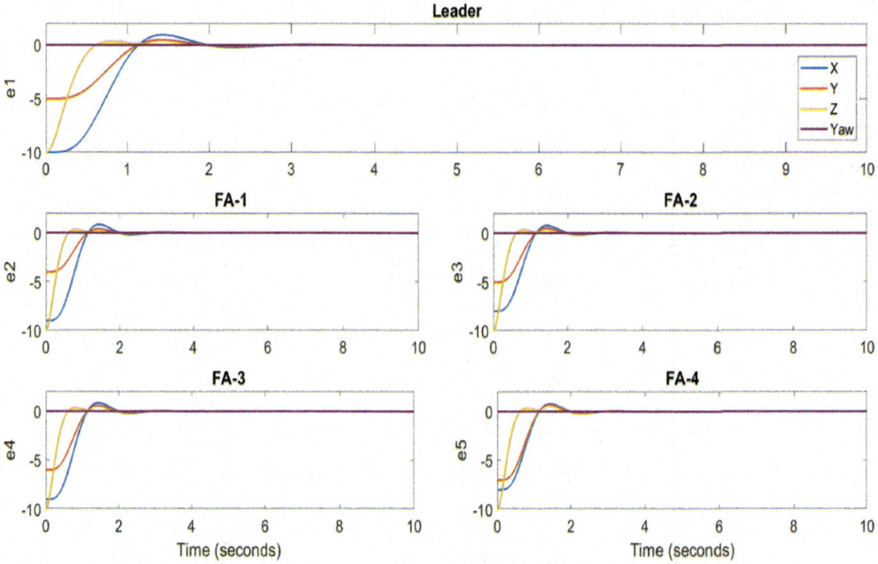

Fig. 6.12 Tracking error of all agents using DDSM Gao's reaching law

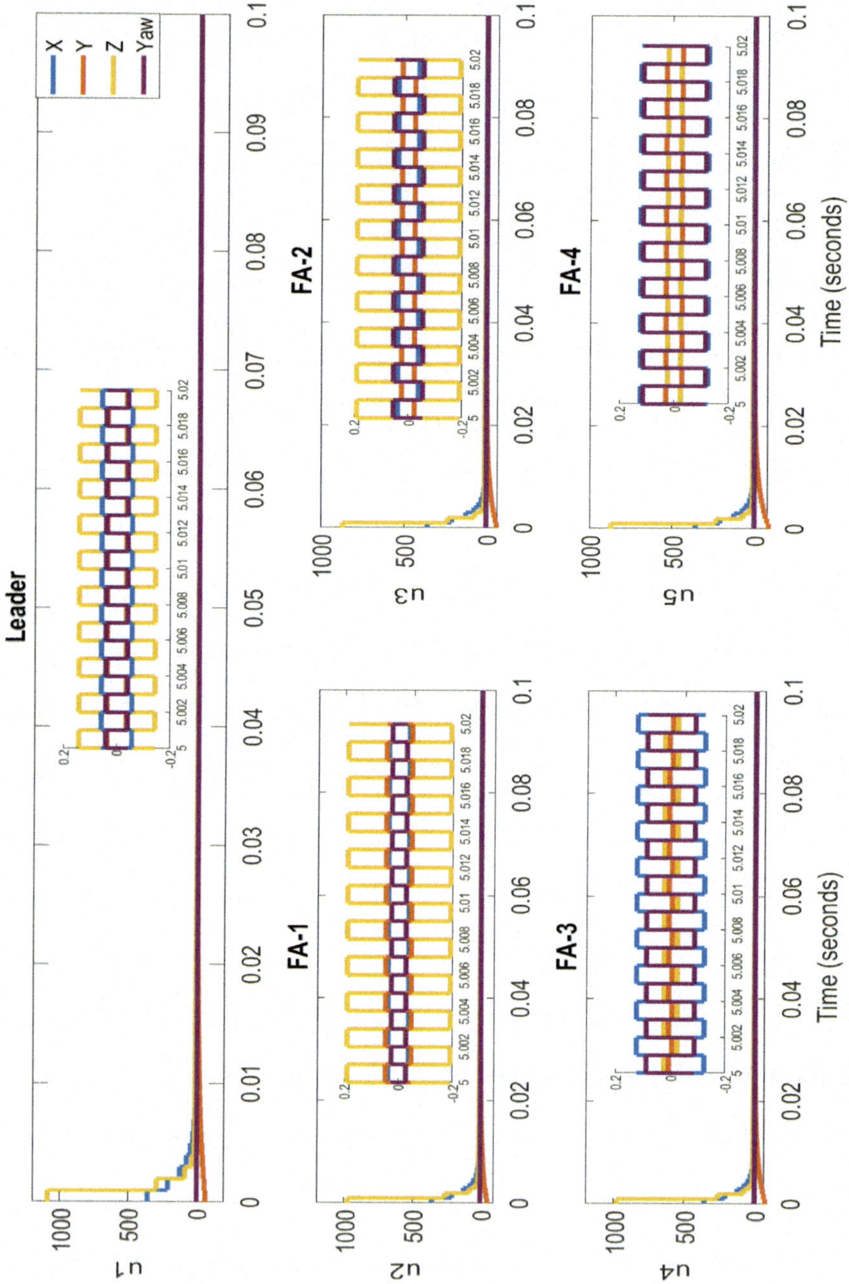

Fig. 6.13 Control efforts of all agents using DDSM Gao's reaching law

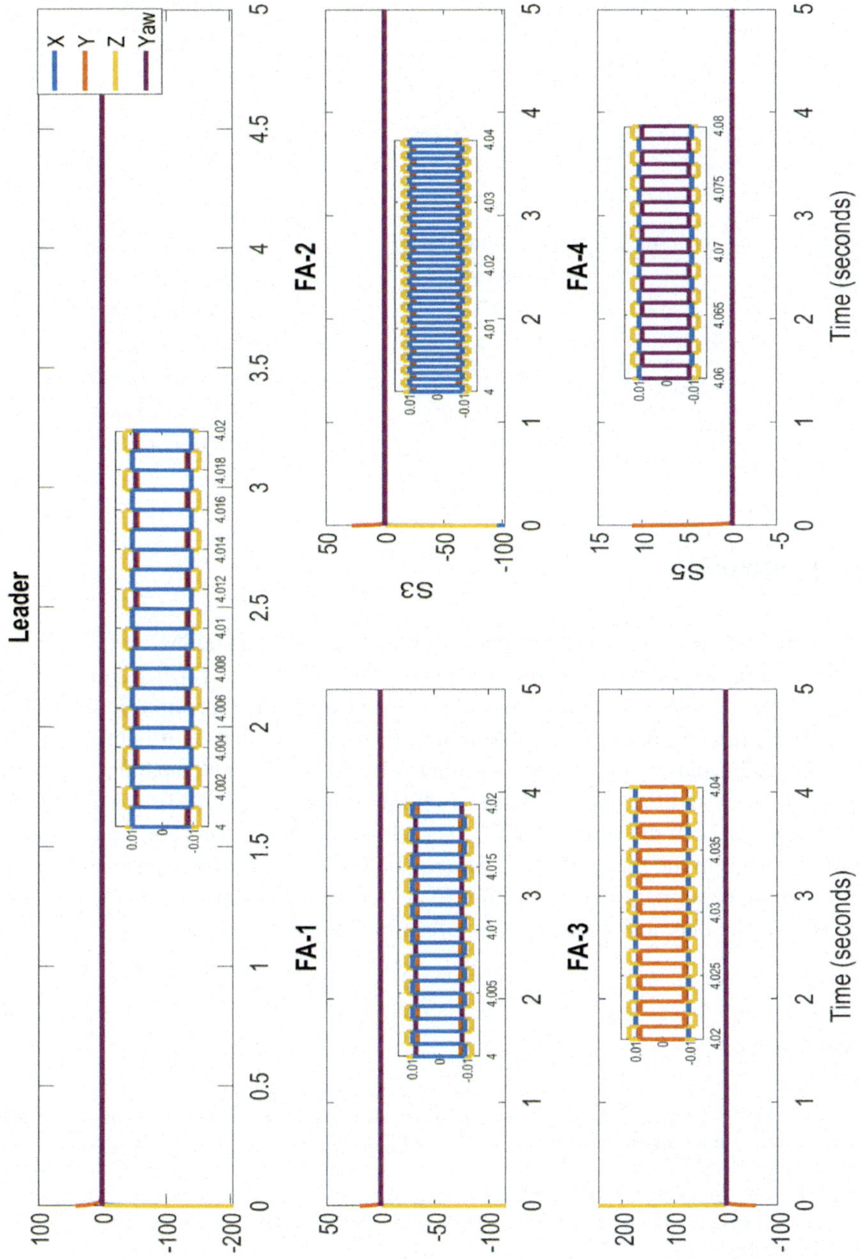

Fig. 6.14 Sliding surface of all agents using DDSM Gao's reaching law

Table 6.1 Comparison study of QSMB (Δ_i)

QSMB	Gao's protocol	Exponential protocol
Δ_1	0.02	2.0×10^{-6}
Δ_2	0.02	2.0×10^{-6}
Δ_3	0.02	2.0×10^{-6}
Δ_4	0.02	2.0×10^{-6}
Δ_5	0.02	2.0×10^{-6}

Table 6.2 Comparison study of control effort

Control effort	Gao's protocol	Exponential protocol
u_1	−0.2 to +0.2	−0.00003 to +0.00003
u_2	−0.2 to +0.2	−0.00001 to +0.00001
u_3	−0.2 to +0.2	−0.00001 to +0.00001
u_4	−0.2 to +0.2	−0.00001 to +0.00001
u_5	−0.2 to +0.2	−0.00001 to +0.00001

6.4 Conclusion

In this chapter, a distributed discrete-time sliding mode (DDSM) protocol was designed using exponential reaching law and DMAS for the formation of Quadcopter swarm in a given pattern. The stability condition of the asymptotic formation is also derived using the Lyapunov function. From the simulation it is inferred that the proposed formation protocol brings all Quadcopters in the given pattern in a finite number of steps and the maximum quasi-sliding mode band (QSMB) as calculated for the protocol is in the range of 10^{-6}. The global control efforts for the formation of swarm varies in the range of 10^{-5}. The robustness of the protocol under the external disturbance and parameter variation is also tested. Also, the proposed protocol outperforms the classical DDSM protocol using Gao's reaching law.

References

1. Han, H., Wu, X., Qiao, J.: Design of robust sliding mode control with adaptive reaching law. IEEE Trans. Syst. Man Cybern.: Syst. **50**(11), 4415–4424 (2020). https://doi.org/10.1109/TSMC.2018.2852626. Nov.
2. Ma, H., Wu, J., Xiong, Z.: A novel exponential reaching law of discrete-time sliding-mode control. IEEE Trans. Ind. Electron. **64**(5), 3840–3850 (2017). https://doi.org/10.1109/TIE.2017.2652390. May

3. Keyurkumar Patel, A.M.: Discrete-time event-triggered higher order sliding mode control for consensus of 2-DOF robotic arms. European J. Control **56**, 231–241 (2020). ISSN 0947-3580, https://doi.org/10.1016/j.ejcon.2020.01.006
4. Gao, W., Wang, Y., Homaifa, A.: Discrete-time variable structure control systems. In: IEEE Transactions on Industrial Electronics, vol. 42, no. 2, pp. 117–122, Apr. 1995. https://doi.org/10.1109/41.370376

Chapter 7
Concluding Remarks and Future Scope

This book presents the robust sliding mode protocols for the formation of Quadcopter swarm in both continuous-time and discrete-time domains. The leader-follower type of Multi-agent system (MAS) concept has been utilized for achieving the formation. In the MAS application, a Quadcopter acts as an individual agent. To establish the communication between those agents, the fixed and undirected graph theory is developed. While using the leader-follower MAS application, there are mainly two types of controllers being used, namely, (1) primary controller and (2) distributed or secondary controller. The primary controller stands for the individual controller of the Quadcopter to get the desired setpoint or to maintain the desired output. However, a distributed controller is used for maintaining the appropriate distance between the neighborhood Quadcopters to achieve the formation as per the given pattern.

Firstly, Chaps. 1 and 2 provide detailed information about the Quadcopter swarm, MAS concept, graph theory, and sliding mode control. Chapter 3 gives information about the distributed continuous-time sliding mode (DCSM) protocol for the formation of linear and non-linear models of a Quadcopter swarm of MAS configured with fixed graph topology for a given pattern. The primary controller for the linear model of the Quadcopter is designed using the LQR technique. Then the first-order constant-proportional reaching law is used to develop the DCSM formation protocol for the linear model of the swarm. As the first-order sliding mode control has a significant chattering effect during the operation, the DCSOSM ST formation protocol is designed to achieve faster convergence and a lesser chattering effect. Both protocols achieve the formation within finite time and the required conditions to ensure the global stability of the formation are derived using the Lyapunov stability function. From the simulation results, it is inferred that the DCSM super-twisting formation protocol has better outcomes than the constant-proportional reaching law.

As the Quadcopter is a highly non-linear system in nature, the DCSM formation protocol is designed for the non-linear model of a Quadcopter swarm. The primary controller is developed using the constant-proportional reaching law type of sliding mode control (SMC) and the distributed controller is designed using the

A. Mehta and A. Modi, *Robust Sliding Mode Protocols for Formation of Quadcopter Swarm*, Studies in Systems, Decision and Control 521,
https://doi.org/10.1007/978-981-99-9726-8_7

super-twisting algorithm. In the non-linear model of the swarm, one predefined trajectory is given as the reference. From the simulation, it is inferred that the proposed formation protocol brings all Quadcopters in the given pattern in a finite time. The robustness of the protocol under external disturbance and parameter variation is also tested.

Nowadays most of the micro-controllers and processors work in digital communication. So, in order to take advantage over continuous-time sliding mode control, discrete-time sliding mode protocols are introduced for the formation of a Quadcopter swarm. In Chap. 4, DDSM protocols are designed for the formation of Quadcopter swarm. The discrete-time primary controller is designed using the LQR technique. Here, a DMAS is used to develop the distributed controller. To begin with, a DDSM formation protocol using the power rate reaching law is designed for the swarm. The performance of the designed protocol using the power rate reaching law is compared with the other first-order controller designed using Gao's reaching law. The analysis of the number of steps required for the global formation of DMAS for both protocols along with the Lyapunov stability analysis is discussed. From the simulation, it is revealed that the controller designed using power rate reaching law has second order ($O(\tau)^2$) quasi-sliding mode band (QSMB) where Gao's reaching law has the first order ($O(\tau)$) QSMB. The robustness of the protocol under external disturbance and parameter variation is also tested. However, the protocol using power rate reaching law improves the performance of DMAS by reducing the QSMB and increases the convergence speed compared to Gao's reaching law.

Further, in Chap. 5 a DDSM super-twisting protocol is designed for the formation of Quadcopter swarm along with the global stability analysis using the Lyapunov function. The discrete super-twisting (DST) protocol not only ensures robustness but also reduces the QSMB and faster the convergence than the first-order DSMC. In order to check the efficacy of the DDSM super-twisting formation protocol, the other two first-order controllers designed using Gao's reaching law and Bartoszewicz's reaching law are compared. The simulation results and the quantitative analysis show that the proposed formation control law outperforms the other controllers.

Chapter 6 discusses the novel DDSM protocol using the exponential reaching law for the formation of Quadcopter swarm. Exponential reaching is also inherently more robust than the other controllers. The number of steps required for the global formation along with the global stability of DMAS is derived using the Lyapunov function. Simulation results of exponential reaching law are compared with Gao's results. It is also reveal that the exponential reaching has the third order ($O(\tau)^3$) QSMB for the sliding surface. Among all the controllers proposed in the book, the controller designed using exponential reaching law has the best results.

❖ Future Scope

Many research studies in the field of distributed control have been performed all over the past decade. However, there are still a lot of challenges that need to be resolved in the field of Quadcopter swarm.

- In this book, a fixed and undirected type of graph has been used for the formation pattern of Quadcopter swarm. So, one can design a formation protocol using the switching as well as a directed type of graph topology so that, the individual Quadcopter can change the formation pattern during the operation.
- In MAS of a Quadcopter swarm, the coordination and communication between the agents are crucial tasks. Hence the Quadcopter is highly non-linear in nature and Because of unpredictable disturbances of environmental factors like temperature, aging equipment from improper maintenance, and battery outages, motor failures, sensor failures, and actuator failures will inevitably occur, ultimately causing the MAS network to a whole to collapse. Considering the uncertain failures, it is important to design the formation protocol using the fault-tolerant protocol for heterogeneous and homogeneous Quadcopter swarms.
- This research work is completely based on the leader-following formation protocol and on the other hand there are some other protocols, such as leaderless, event-triggered, bipartite formation, etc., that can be configured for the homogeneous and heterogeneous system.

The manufacturer's authorised representative in the EU is Springer Nature Customer Service Centre GmbH, Europaplatz 3, 69115 Heidelberg, Germany. If you have any concerns regarding our products, please contact ProductSafety@springernature.com

Printed and bound by CPI Group (UK) Ltd, Croydon, CR0 4YY
01/10/2026
02225883-0011